The Spirit of Mathematics

THE SPIRIT OF MATHEMATICS

algebra and all that

DAVID ACHESON

OXFORD
UNIVERSITY PRESS

OXFORD
UNIVERSITY PRESS

Great Clarendon Street, Oxford, OX2 6DP,
United Kingdom

Oxford University Press is a department of the University of Oxford.
It furthers the University's objective of excellence in research, scholarship,
and education by publishing worldwide. Oxford is a registered trade mark of
Oxford University Press in the UK and in certain other countries

Published in the United States of America by Oxford University Press
198 Madison Avenue, New York, NY 10016, United States of America

British Library Cataloguing in Publication Data
Data available

Library of Congress Control Number: 2022941432

ISBN 978–0–19–284508–5

Contents

1
Introduction

Is it possible to capture the whole spirit of mathematics at its best *using only simple materials*?

I certainly hope so, because that is what this book is all about. And the 'simple materials' that I have in mind are just the most elementary parts of arithmetic, algebra, and geometry that we all meet at school.

"THIS IS THE PART I ALWAYS HATE."

Fig. 1 Mathematics in action

For many people, I think, the most mysterious of these is *algebra*.

And, so far as I can determine, the mystery can often be summed up by one simple question: what is algebra really *for*?

* * *

First and foremost, in my view, algebra helps us express *general* statements and ideas in mathematics. And this is illustrated most simply, I think, by the notion of a formula.

Fig. 2, for instance, shows the formula for the vibration frequency of a guitar string. And we will see in due course not only what all the symbols mean, but how the formula actually works in practice.

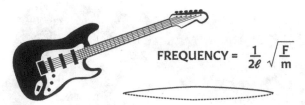

$$\text{FREQUENCY} = \frac{1}{2\ell} \sqrt{\frac{F}{m}}$$

Fig. 2 Vibrations of a guitar string. Here F denotes the tension force in the string, l its length, and m its mass per unit length

* * *

A quite different kind of generality in algebra is shown by the result in Fig. 3, which is, instead, of a purely mathematical kind, and holds for *any* numbers x and a, positive or negative.

$$(x+a)^2 = x^2 + 2ax + a^2$$

Fig. 3 Algebra at its best

And, believe it or not, this particular result will turn out to be something of a 'star' in this book. In due course, then, I will again want to explain what the symbols mean, and why the result itself is true.

* * *

At this point, however, I suspect that some readers may be a little puzzled, and quietly thinking to themselves: 'I thought algebra was all about *finding x*'.

And it is certainly true that determining some initially unknown number has always played a major part in the history of the subject.

3. Find x.

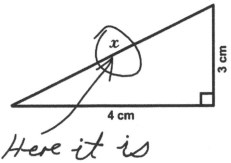

Fig. 4 A spoof—or a genuine piece of school work? No one seems to know

Many of us, after all, make a start on algebra with little problems like this:

> In 7 years' time I will be twice as old as I was 7 years ago.
> How old am I now?

If we let x denote my present age (in years), we are told, then, in effect, that

$$x + 7 = 2\,(x - 7).$$

Multiplying out the brackets on the right-hand side gives

$$x + 7 = 2x - 14,$$

and subtracting x from both sides then tells us that $x - 14 = 7$.

So I am 21. (In my dreams.)

And the only drawback with little problems like this is that they can seem rather artificial or contrived.

Or, even, downright ridiculous…

2
Whatever Happened to A, B, and C?

When I was at school in the 1950s, we did lots of problems like this:

> A and B, working together, can fill a bath in 4 hours. A and C, working together, can fill the same bath in 5 hours. B can fill twice as fast as C.
>
> How long would C take to fill the bath, working alone?

Fig. 5 Mathematical bath-filling

Now, the trick with this sort of problem is always to focus on what *fraction* of the task each person can accomplish in a given time—say 1 hour.

So, if we let these fractions be *a*, *b*, and *c*, then we know that

$$a+b=\frac{1}{4},$$

$$a+c=\frac{1}{5},$$

$$b=2c,$$

because A and B together can fill $\frac{1}{4}$ of the bath in 1 hour, and so on.

In this way, then, we have three equations for the three 'unknowns' *a*, *b*, and *c*.

Now, it is really only the value of *c* that we are interested in, so it makes sense to try and get rid of both *a* and *b*.

Getting rid of *b* is easy; we can use the third equation $b = 2c$ to substitute for *b* in the first, giving

$$a + 2c = \frac{1}{4}.$$

Then, if we rewrite the second equation as $a = \frac{1}{5} - c$, we can use that to get rid of *a*:

$$\frac{1}{5}-c + 2c = \frac{1}{4},$$

which simplifies to

$$c=\frac{1}{4}-\frac{1}{5}=\frac{1}{20}.$$

So C can fill $\frac{1}{20}$ of the bath in 1 hour, and would therefore take 20 hours to fill the bath on his own, which—frankly—doesn't bear thinking about.

The human element in mathematics?

It is easy to poke fun at such problems, though not perhaps as skilfully as in a famous essay by the Canadian humourist Stephen Leacock, first published in 1910.

Leacock claims that A is always the strongest and most energetic of the three, B comes next, and C is the weakling.

Fig. 6 Stephen Leacock (1869–1944), author of 'A, B, and C; the Human Element in Mathematics'

For, according to Leacock:

Poor C is an undersized, frail man, with a plaintive face. Constant walking, digging and pumping has broken his health and ruined his nervous system...as Hamlin Smith has said, 'A can do more work in one hour than C in four.'

There is, however, some evidence that Leacock did not do his research as thoroughly as one might hope.

For James Hamblin Smith, to whom Leacock refers (slightly inaccurately), was a highly successful private tutor at Cambridge in the nineteenth century, and if we turn to page 172 of his *Treatise on Arithmetic* (1889) we do indeed find A, B, and C engaged in a variety of tasks *including the one in Fig. 7.*

(6) *A* does a piece of work in 3 hours, which is twice the time *B* and *C* together take to do it; *A* and *C* could together do it in 1⅓ hours.   How long would *B* alone take to do it?

Fig. 7 From Hamblin Smith's *Arithmetic* of 1889

Now, there is something very odd going on here.

We are told that A can do the work in 3 hours, so imagine for a moment that A has a twin brother. Together, they will do the work in $1\frac{1}{2}$ hours. Yet, working with C, A can do it more quickly—in just $1\frac{1}{3}$ hours.

This means that C must be *working faster than* A!

Further investigation reveals that C works faster than B, as well. It is truly C's finest hour.

And, no less remarkably, B eventually goes mad (Fig. 8).

(468) *A* can do a piece of work in 6 days, which *B* can destroy in 4.   *A* has worked for 10 days, during the last 5 of which *B* has been destroying : how many days must *A* now work alone, in order to complete his task?

Fig. 8 An unexpected twist, from Hamblin Smith's *Arithmetic* of 1889

Stylish bath-filling?

Strange as it may seem, even bath-filling offers opportunities for elegant mathematics.

Consider, for instance, this problem:

> A and B can fill a bath in 3 hours; A and C in 4 hours; B and C in 6 hours. In what time would they fill it, all working together?

We begin, as before, by letting a, b, and c be the fractions of the bath that each can fill in 1 hour, working alone. This gives us the three equations:

$$a + b = \frac{1}{3},$$

$$a + c = \frac{1}{4},$$

$$b + c = \frac{1}{6}.$$

Now, we could solve these equations, by the same sort of method as before, to find the value of c, and once we know that, the values of a and b will follow quite quickly, from the second and third equations.

It is worth noting, however, that these individual values of a, b, and c simply do not matter so far as this particular problem is concerned; all we really need to know to get the 'all working together' answer is the *sum* $a + b + c$.

And there is a neater way of getting this.

We simply notice that

$$2(a+b+c) = (a+b)+(a+c)+(b+c)$$

$$= \frac{1}{3}+\frac{1}{4}+\frac{1}{6},$$

which gives us the value of $a+b+c$ straight away as $3/8$, leading to a time of $8/3 = 2\,2/3$ hours to fill the bath, all working together.

And it seems to me, at least, that even in such an unpromising and fanciful setting we have here a perfectly good example of elegant mathematics using only simple materials.

3
The 1089 Trick

Mathematics at its best often has an element of *surprise*, and the simplest example I know is a magic trick.

The first step is to write down a three-figure number.

Any such number will do, as long as the first figure is greater than the last by 2 or more.

Now reverse your number and subtract. Finally, reverse the result and add.

The final answer is then *always* 1089.

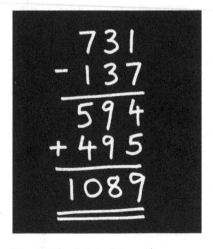

Fig. 9 The 1089 trick

Fig. 10 An exciting moment, from long ago

I first came across this at the age of 10, in the *I-SPY Annual* for 1956 (Fig. 10).

And while it is not exactly 'serious' mathematics, it blew my socks off.

Why does it work?

Let the starting number have digits a, b, and c, where $a - c$ will be greater than 1. Then that number is actually $100a + 10b + c$, and after reversing and subtracting we have

$$100a + 10b + c - (100c + 10b + a)$$
$$= 100a + 10b + c - 100c - 10b - a$$
$$= 99a - 99c$$
$$= 99(a - c).$$

So, at the end of the first part of the trick, *we always have a multiple of 99.*

Now, $a - c$ is at least 2 and at most 9, so the possible multiples of 99 are

$$198$$
$$297$$
$$396$$
$$495$$
$$594$$
$$693$$
$$792$$
$$891,$$

and as we proceed down the list, the first digits increase by 1 each time, while the last digits decrease by 1.

There is no mystery to this of course; adding an extra 99 is equivalent to adding 100 and subtracting 1. In consequence, the first and last digits of any number in the list *always add up to 9.*

So, when we reverse any one of these numbers and add—which is the last part of the 'trick'—we get 9 lots of 100 from the first digits, 9 lots of 1 from the third digits, and 2 lots of 90 from the second, giving

$$900 + 9 + 180 = 1089.$$

Who invented it?

The history of this trick is a little curious.

In the *Boy's Own Paper* for 1893, a version of it appeared using pounds, shillings, and pence!

The final answer is then always £12 18s 11d,

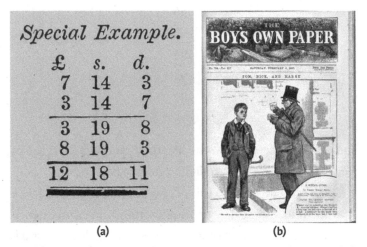

(a)   (b)

Fig. 11 From the *Boy's Own Paper* (1893). Here 1 pound (£) is 20 shillings, and 1 shilling (s) is 12 pence (d)

One intriguing possibility is that this version of the trick was invented by the Oxford mathematician Charles Dodgson—better known as Lewis Carroll, author of *Alice in Wonderland*—though the only evidence for this appears to come from *The Lewis Carroll Picture Book* of 1899. This was compiled by his nephew, who describes the trick as a

'numerical curiosity which I believe to have been discovered by Mr Dodgson'.

In any event, the monetary version seems to have preceded the 1089 trick itself, and predominated until the early 1950s, reappearing from time to time in books such as *Willane's Wizardry* and *Uncle Mac's Children's Hour Story Book*.

Another explanation

The earliest appearance of the 1089 trick *itself* seems to be in an 1896 French edition of a (now classic) book on mathematical recreations by Rouse Ball, and it features a slightly different explanation of why the trick works (Fig. 12).

ARITHMÉTIQUE		15

Le tableau récapitulatif suivant explique avec un exemple notre règle.

(1°)	732	$100.a + 10.b + c$	On suppose $c < a.$
(2°)	237	$100.c + 10.b + a$	
(3°)	495	$100(a - c - 1) + 90 + (10 + c - a)$	
(4°)	594	$100(10 + c - a) + 90 + (a - c - 1)$	
(5°)	1089	$900 + 180 + 9$	

Fig. 12 The 1089 trick, in *Recreations et Problemes Mathematiques des Temps Anciens et Modernes*, by J. FitzPatrick (1896). At stage 3, in order to avoid the second bracket being negative, one of the 100s has been converted into $90 + 10$

I myself have always preferred the first explanation above, which strikes me as more elementary. But I can well understand why some might prefer the one in Fig.12, on account of the neat way in which a, b, and c *all* cancel out!

4
Another Kind of Magic

One of the pleasures of mathematics at its best lies in some of the actual methods used, and I have always rather liked the whole idea of *proof by contradiction*.

Here, the aim is to prove some proposition by showing that if it were *not* true, some kind of contradiction or absurdity would follow.

And the only drawback with this method of proof is that you can't usually tell in advance *how* the contradiction or absurdity is going to arise.

So you have to keep your wits about you!

"It is an old maxim of mine that when you have excluded the impossible, whatever remains, however improbable, must be the truth."

Fig. 13 Proof by contradiction? Or just something a bit like it? This famous Sherlock Holmes quote appears in 'The Adventure of the Beryl Coronet'

And to illustrate this I would now like to turn to one of the best opportunities I know for elegant mathematics with simple materials.

A magic square

This is a square array of numbers such that all the rows, all the columns, and both diagonals add up to the same total, called the 'magic constant', M.

Fig. 14 shows the simplest example, using just the first 9 whole numbers.

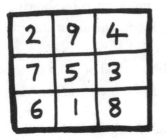

Fig. 14 A magic square

But how is such a square actually constructed?

The magic constant

The first step is to calculate the magic constant M, and this is quite easy because the numbers $1,\ldots,9$ add up to 45, and there are 3 rows, each with total M.

So M must be 15.

What goes in the central box?

The answer is 5, and we can prove this by contradiction.

Suppose first that the number in the central box is less than 5: 4, for example.

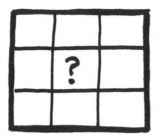

Fig. 15 What goes in the centre?

The question then arises: *where is the 1 going to go?* Wherever we put it, we find ourselves with some row, column, or diagonal in which the most we can manage is 1 + 4 + 9 = 14, rather than 15.

The same kind of problem arises if the number in the central box is 3, 2, or 1.

So the number in the central box cannot be less than 5.

And if we put a number greater than 5 into the central box—6, say—we run into an equivalent problem: where is the 9 going to go?

So the number in the central box must be 5.

9 can't go in a corner box

This, too, we can prove by contradiction.

Assume, then, that 9 can go in a corner (Fig. 16), in which case the number in the opposite corner will have to be 1.

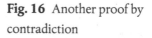

Fig. 16 Another proof by contradiction

But that leaves us needing both A + B = 6 and C + D = 6 in Fig. 16, and this can't be done, because with the 1 and 5 'gone' there are only three numbers less than 6 left, namely 2, 3, and 4.

So 9 can't go in a corner.

And so to the end...

Having put 9 in a non-corner position (Fig. 17a), we now need E + F = 6, with only 2, 3, and 4 to play with, which leads to two possibilities:

$$E = 2, F = 4 \ or \ E = 4, F = 2.$$

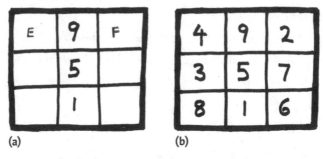

(a)   (b)

Fig. 17 The final stages

With the help of the magic constant 15, the first alternative then leads to our original magic square in Fig. 14, while the second leads to Fig 17b, which is essentially the same magic square but viewed from 'round the back'.

In fact, there is really only one magic square involving just the numbers 1, 2,..., 9; all the others are 'trivial' rearrangements—such as rotations—of the one in Fig. 14.

* * *

But while the preceding arguments do demonstrate the idea of proof by contradiction, they don't really convey its full power.

This is because we used it, on each occasion, to reject just a small handful of other possibilities.

Arguably, however, the full power of proof by contradiction only really kicks in when the number of other possibilities to be rejected is *infinite*.

Before x

c. 1800 BC

Algebra problems, *in word form*, on Babylonian clay tablets.

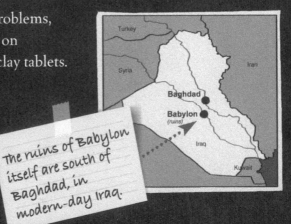

The ruins of Babylon itself are south of Baghdad, in modern-day Iraq.

c. 850 AD

Al-Kharizmi was a Persian mathematician at the 'House of Wisdom' in Baghdad.

His famous treatise *Al-jabr w'al Muqabala* gave 'algebra' its name.

1557

Robert Recorde introduces the 'equals' sign in his *Whetstone of Witte*, in an equation we would write as 14x + 15 = 71:

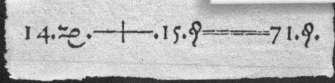

1637

René Descartes, in his *La Géométrie*, introduces a, b, c, etc. for 'given' numbers and x, y, z, etc. for 'unknowns' that we are trying to find.

5

Just Imagine…

Mathematics often requires a bit of *imagination*, and many good examples can be found in elementary geometry.

To see what I mean, first take a compass, draw a circle, and then draw in a diameter, with end points A and B (Fig. 18).

Finally, pick any point P on the circumference and join it to A and B by straight lines.

Then the angle at P is *always* 90°.

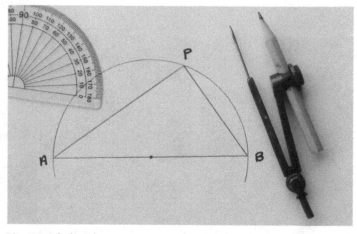

Fig. 18 Thales' theorem

This surprising result dates from ancient Greece in around 600 BC, and is known as *Thales' theorem*.

And proving it requires a little imagination, as we will see very shortly.

Euclid's *Elements*

While Thales is sometimes credited with introducing geometry as a carefully ordered and deductive subject, it was really Euclid who took the whole idea of theorem and proof to new and dizzy heights, some 300 years later.

In spite of its ultra-concise style of exposition (or even, perhaps, because of it), Euclid's *Elements* has had more influence—and has run to more editions—than virtually any other book in human history.

Fig. 19 A popular edition of Euclid's *Elements*, from the original (in French) by de Chasles

In the nineteenth century, for instance, there was one headmaster at Eton of whom it used to be said:

> He divided the books of the world into three classes:
> Class I: The Bible
> Class II: Euclid
> Class III: All the rest.

And in order to prove Thales' theorem we will need two results about triangles which Euclid proves early on.

First, if we have an *isosceles* triangle, i.e. one with two equal sides, then its so-called 'base angles' are always equal (Fig. 20).

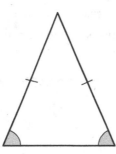

Fig. 20 An isosceles triangle

Many people, I suspect, find this particular result rather obvious. After all, if we simply 'flip over' the triangle in Fig. 20, it will look exactly the same.

The other result that we need is, I think, far less obvious, though quite well known: *the three angles of any triangle add up to 180°.*

In this connection, many readers will probably be familiar with the little experiment in Fig. 21, where we tear up a paper triangle and reposition the pieces to form an apparently straight line, and I think it is worth being clear why this is *not* a proof. It is not simply because of the inevitable 'experimental error', but—far more fundamentally—because each time the experiment is done, it relates only to *one particular triangle*.

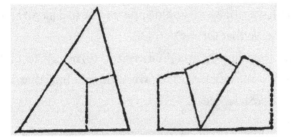

Fig. 21 The angle-sum of a triangle, from W. D. Cooley's *Elements of Geometry, Simplified and Explained* (1860)

Yet the result is entirely *general*, and just what we need, in fact, for the task ahead...

Proof of Thales' theorem

We want to prove that if P is any point on the semicircle in Fig. 18, then angle APB = 90°.

Now, in my experience at least, we hardly ever prove anything in geometry by just staring at the original diagram; we almost always have to *do something to it*. In other words, we have to play around with the problem and experiment a bit.

So, imagine for a moment taking the point P *off* the semicircle and moving it far away from A and B. It seems obvious to me, at least, that we can then make the angle APB very small indeed. Similarly, by moving P sufficiently close to the centre of the circle O, we can make angle APB as close as we like to 180°.

So the result plainly isn't true if P is just *anywhere*, and we must therefore find a way of expressing, mathematically, that P lies on the semicircle.

Now, the defining property of a circle is that all its points are the same distance from its centre O, so the most down-to-earth way forward would appear to be to *draw in the line* OP, and observe that OP = OA = OB.

And this slightly imaginative step turns out to be something of a masterstroke, for we suddenly find that we have two *isosceles* triangles.

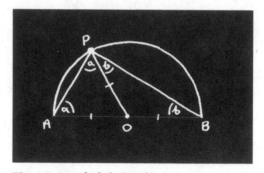

Fig. 22 Proof of Thales' theorem

So, in Fig. 22, the two 'base angles' *a* are equal, and so also are the other two base angles *b*.

Finally, the three angles of the original triangle APB must add up to 180°, so

$$a + (a + b) + b = 180°.$$

Therefore $2(a + b) = 180°$, and, in consequence, $(a + b) = 90°$. So angle APB = 90°, which proves the theorem!

6

A Most Unusual Lecture

In October 1903, Frank Nelson Cole, a professor at Columbia University, gave a very strange lecture to the American Mathematical Society.

Cole was a man of few words at the best of times, but on this particular occasion he gave his whole presentation *without saying anything at all.*

Fig. 23 Frank Nelson Cole (1861–1926)

Instead, he wrote the number

$$2^{67} - 1$$

on a blackboard, and proceeded to multiply it out, by hand, until he eventually obtained

$$147, 573, 952, 589, 676, 412, 927.$$

Then, on another blackboard, he wrote down the following product:

$$193, 707, 721 \times 761, 838, 257, 287$$

and worked that out, again by hand, showing—eventually—that it gave *the same final answer*.

Then he sat down.

And in order to fathom what was really going on here, we need to venture briefly into the slightly weird world of *prime numbers*.

Prime time

A prime number is a positive whole number which is divisible only by itself and 1.

Thus 13 is prime, for example, but 15 is not, because it is divisible by 3 or 5.

The first few primes are

$$2, 3, 5, 7, 11, 13, 17, 19, 23\ldots$$

and any positive whole number is either prime *or* can be written as the product of two or more prime factors (Fig. 24).

Fig. 24 Prime factors

Now, numbers of the form

$$2^n - 1$$

play a special part in the theory, and are named after the French monk Marin Mersenne (1588–1648).

A Mersenne number can only be prime if n itself is prime, and the first few examples *are* prime:

$$2^2 - 1 = 3$$
$$2^3 - 1 = 7$$
$$2^5 - 1 = 31$$
$$2^7 - 1 = 127$$

Yet the next prime value of n delivers a Mersenne number which is not prime:

$$2^{11} - 1 = 2047$$
$$= 23 \times 89.$$

So n being prime is necessary for $2^n - 1$ to be prime, but not sufficient.

And, as it happens, when Cole approached the blackboard in 1903 it had long been known that $2^{67} - 1$ is *not* prime. The trouble was...the prime factors were so large that no one had been able to find them!

Indeed, actually finding the prime factors of some truly enormous number can be all but impossible even today, and, as many readers will probably know, this is the basis of public-key cryptography and internet security.

And Mersenne primes continue to play a major part in the theory, for at the time of writing the largest known prime is of Mersenne type:

$$2^{82,589,933} - 1,$$

with over 24 million digits.

Yet there is no danger of prime numbers ever 'running out'.

This is because of an extraordinary result, from over 2,000 years ago, showing that the number of primes is *infinite*.

Infinitely many primes

We will prove this by contradiction, using a slight variation on the original proof by Euclid of Alexandria.

Fig. 25 Euclid of Alexandria

Suppose that the number of primes is *finite*.

There will then be some *largest* prime number, which we will call p.

Now consider the number obtained by multiplying all the primes together, and adding 1:

$$N = 2 \times 3 \times 5 \times \ldots \times p + 1.$$

This number is certainly greater than p, and as p is the largest prime this new number N cannot be prime. It must therefore be possible to write it as a product of primes, i.e. it must be divisible by at least one prime number.

But it isn't, because of the way we've constructed it; if you divide N by any prime number from the (supposedly complete) list 2, 3, 5,..., p you always get a remainder of 1.

We have arrived at a contradiction, then, and the only way out is for the original hypothesis to be wrong.

So the number of primes is infinite.

7

Why Are Mathematicians Obsessed by Proof?

Some problems in mathematics can be very easy to state, yet incredibly hard to solve.

In 1886, for example, J. M. Wilson, headmaster at Clifton College in Bristol, posed the following 'Challenge Problem' to the whole school:

> Prove that it is possible to colour any map in the plane (with neighbouring countries coloured differently) using at most four colours.

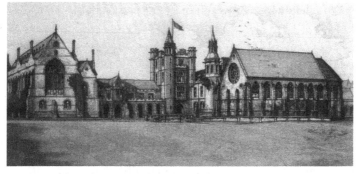

Fig. 26 Clifton College, Bristol in 1898

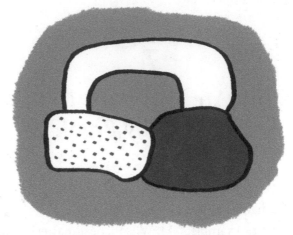

Fig. 27 A simple map where four colours are needed

The problem had originated some 30 years earlier, and Fig. 27 shows a simple example where four colours *are* needed.

Wilson asked for solutions from his young pupils 'on or before Dec. 1' but with one strict condition attached:

> No solution may exceed one page, 30 lines of MS., and one page of diagrams.

But this turned out to be spectacularly optimistic.

When the four-colour theorem was finally proved, *90 years later*, by K. Appel and W. Haken, the proof contained 10,000 diagrams, and the associated computer output stood four feet high on the floor. It even raised fundamental questions about the whole notion of proof in mathematics.

'Try to prove theorem'

For many people, I think, the key question about proof is a rather more down-to-earth one, namely: why are mathematicians so obsessed with proof *at all*?

The most obvious answer, of course, is that without proof the proposition in question might simply be *wrong* (Fig. 28).

Monday:	Try to prove theorem.
Tuesday:	Try to prove theorem.
Wednesday:	Try to prove theorem.
Thursday:	Try to prove theorem.
Friday:	Theorem false.

Fig. 28 This slightly tongue-in-cheek view of a typical research mathematician's week is due to Julia Robinson (1919–1985)

But if we probe a little more deeply, and ask how this comes about, we find that it is largely because mathematicians love making *general* statements, claiming that such-and-such is true for an infinite number of special cases.

And this is, in a sense, living dangerously.

A tempting proposition

Consider, for instance, the following:

> Proposition: For all positive whole numbers n, $991n^2 + 1$ is
> *not* a perfect square.

Now, by 'perfect square' I mean here the square of a whole number, and unless you happen to be an expert in this area of mathematics (I am most certainly not) it is natural to start by considering a few simple cases.

If $n = 1$, for instance, $991n^2 + 1 = 992$, which is not a perfect square, the nearest being $31^2 = 961$.

If $n = 2$, we find that $991n^2 + 1 = 3965$, which is something of a 'near miss', because $63^2 = 3969$.

Nonetheless, $n = 2$ does *not* lead to a perfect square, and we could, in principle, use a computer to keep checking in this way until, say, $n = 1000$.

If a thousand special cases doesn't strike you as quite enough, what about a million?

Or even, say, a million million?

As it happens, even this many checks is nowhere near good enough.

The proposition is simply *false*, but the first value of n at which we learn that it is false is

$$n = 12,055,735,790,331,359,447,442,538,767.$$

And that's why mathematicians need *proof*.

8

Puzzling Mathematics

This book is about mathematics with simple materials, and before going any further I would like to explore briefly what can sometimes be done *with no materials at all.*

To this end, here are five puzzle-book problems which all contain elements of what I would call 'mathematical thinking' (solutions on p. 156).

A chocolate bar problem

The chocolate bar in Fig. 29 is marked into 6 × 4 = 24 little squares, and we wish to break up the bar into those 24 squares.

At any intermediate stage, we may pick up one piece and break it along one of its marked vertical or horizontal lines.

Fig. 29 Breaking chocolate

And the question is: what is the smallest number of breaks needed?

Without wishing to spoil the problem, it seems to me that it illustrates well one particular aspect of mathematics at its best, namely getting to the heart of some matter by *identifying and dismissing what's irrelevant*.

The roll of the dice

A dice is rolled, without slipping, along the track in Fig. 30, turning two corners. We are reminded that opposite faces of the dice add up to 7, and the question is:

> What number will be showing on the top face when the dice reaches the end of the track?

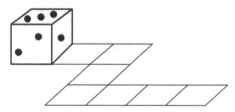

Fig. 30 A dice problem

This problem gives scope, I feel, for an *imaginative* solution, in which we focus throughout only on the one face that is going to matter.

C 'wins' again!

My next example is a slightly unusual problem featuring A, B, and C, which appears in John Bonnycastle's *Scholar's Guide to Arithmetic* of 1806:

There is an island 73 miles in circumference, and 3 footmen all start together to travel the same way about it; A goes 5 miles a day, B 8, and C 10; when will they all come together again?

As it happens, however, I first came upon this problem in an entirely different way.

Fig. 31 From William Botham's exercise book, 1819

Some years ago, I acquired a handwritten exercise book of arithmetic, dated 1819, that once belonged to someone called William Botham (Fig. 31).

Sadly, there is no indication of who this was, or where he lived, and after 200 years the book itself is rather falling apart.

But William Botham's attempt at the unusual A, B, and C problem really caught my eye, because he gets the right answer, but only after four whole pages filled with excruciating long divisions like the one in Fig. 31.

So, once you have recovered from the shock of C *walking faster than* A *or* B, can you, perhaps, do a bit better?

The mutilated chessboard

This puzzle is, I think, rather well known, but it offers a nice opportunity for 'proof by contradiction'.

Fig. 32 shows a chessboard with two opposite corners removed, leaving 62 squares, and we are given 31 dominoes, each able to cover two adjacent squares.

Fig. 32 The mutilated chessboard

Yet it is impossible to cover all the 62 remaining squares with these 31 dominoes.

Why?

A four-card puzzle

In mathematics, at any level, it is always important to distinguish between a statement

P implies Q

and its *converse*

Q implies P,

which, like the original statement itself, may or may not be true.

Suppose, for instance, that we are shown four cards (Fig. 33), and told that each has a number on one side and is either black or white on the other.

Fig. 33 A four-card puzzle

It is then claimed that

If a card is black on one side, then it has an even number on the other.

And the question is: which cards do we need to turn over to determine whether this claim is true or false?

This last example is, in fact, a famous test that is well known to cognitive psychologists.

And lots of people get the answer wrong!

'Quite Enthralling'

BATH-FILLING in mathematics has a long history.

Filippo Calandri's arithmetic book of 1518 features a cistern which can be filled by a pipe in 4 days and emptied by an outlet in 11 days.

How long will it take to fill with the outlet open?

According to her autobiography, the great crime-writer Agatha Christie studied arithmetic as a child up to 'the point where… tanks filled with water in so many hours. I found it quite enthralling.'

9
Why Does $(-1) \times (-1) = +1$?

Most of us, I think, find $(-1) \times (-1) = +1$ very puzzling when we first meet it.

In Daniel Fenning's *Young Algebraist's Companion* of 1750, for instance (Fig. 35), the student asks of his master:

> I am not yet fully satisfied…that $- \times -$ should produce +;
> have you no other way to demonstrate it but barely telling
> me so?

And in order to find the answer, we need to begin by considering a little more carefully the whole foundations of algebra itself, which are based on five key rules.

Addition

Rules 1 and 2 say, loosely speaking, that the order in which we add numbers doesn't matter (Fig. 34).

$$1. \quad a + b = b + a$$
$$2. \quad a + (b + c) = (a + b) + c$$

Fig. 34 Addition rules

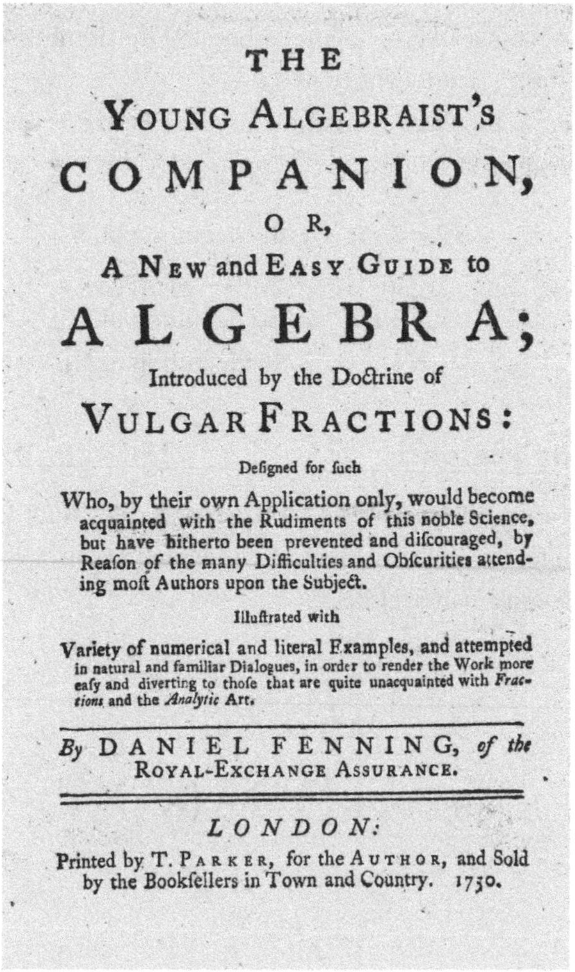

THE
YOUNG ALGEBRAIST'S
COMPANION,
OR,
A NEW and EASY GUIDE to
ALGEBRA;
Introduced by the Doctrine of
VULGAR FRACTIONS:

Designed for such

Who, by their own Application only, would become acquainted with the Rudiments of this noble Science, but have hitherto been prevented and discouraged, by Reason of the many Difficulties and Obscurities attending most Authors upon the Subject.

Illustrated with

Variety of numerical and literal Examples, and attempted in natural and familiar Dialogues, in order to render the Work more easy and diverting to those that are quite unacquainted with *Fractions* and the *Analytic Art.*

By DANIEL FENNING, *of the* ROYAL-EXCHANGE ASSURANCE.

LONDON:
Printed by T. PARKER, for the AUTHOR, and Sold by the Booksellers in Town and Country. 1750.

Fig. 35 *The Young Algebraist's Companion, 1750*

And here's an example of that idea in action:

A proof that 2 + 2 = 4.

This is not a joke, incidentally, because the number 4 is not *defined* as 2 + 2; it is defined as 3 + 1.

There *is*, therefore, something to prove here, and one careful way of proceeding is as follows:

$$2 + 2 = 2 + (1 + 1) \text{ (by definition of 2)}$$
$$= (2 + 1) + 1 \text{ (by rule 2)}$$
$$= 3 + 1 \quad\quad \text{ (by definition of 3)}$$
$$= 4 \quad\quad\quad \text{ (by definition of 4)}$$

Multiplication

Once again there are two key rules (Fig. 36). And, loosely speaking, they say that the order in which we *multiply* numbers doesn't matter, either.

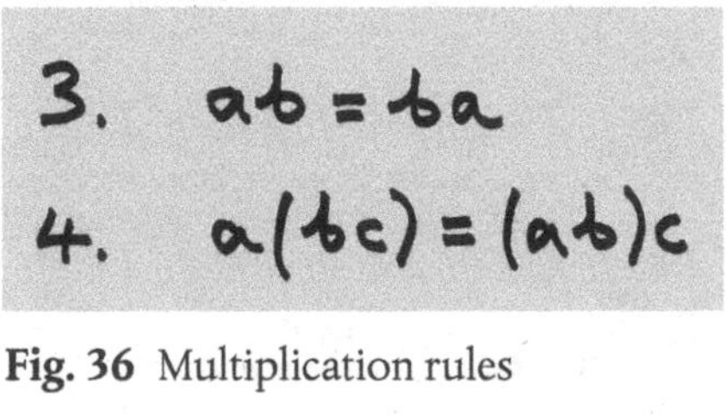

3. $ab = ba$

4. $a(bc) = (ab)c$

Fig. 36 Multiplication rules

(Here, as usual in algebra, the multiplication signs are omitted, so that *ab* is shorthand for the product $a \times b$.)

Area

This must surely be one of the oldest examples of multiplication in action.

We begin with a square of side 1 unit, and it quickly becomes evident how to calculate the area A of a rectangle with sides which are whole numbers (Fig. 37).

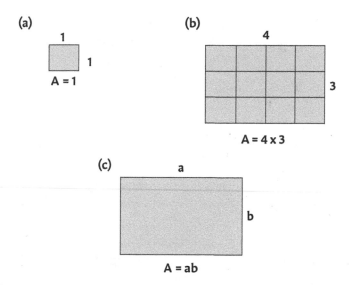

Fig. 37 Area

And it clearly doesn't matter whether we view the area in Fig. 37b as being composed of 3 lots of 4 units or 4 lots of 3.

More generally, indeed, we are led to define the area of any rectangle as $A = ab$ (Fig. 37c), where the side lengths a and b

may now be any numbers, not necessarily whole numbers or even ratios of whole numbers.

The distributive rule

This rule is a little different, because it involves ideas of both addition *and* multiplication (Fig. 38).

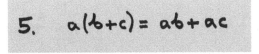

Fig. 38 The distributive rule

Applying Rule 3 (three times!) produces the following alternative form of Rule 5:

$$(b + c)\, a = ba + ca,$$

which is just as useful in practice.

Once again, both make sense in terms of area, because the area of the large rectangle in Fig. 39 is plainly the sum of the areas of the two smaller ones.

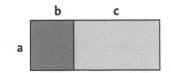

Fig. 39 Illustrating the distributive rule

Negative numbers

In the examples so far, all the numbers have been positive.

Yet all five rules hold, expressly, for all numbers a, b, c, positive or negative.

If, for instance, I go into a shop and want to buy 7 pairs of socks at 99p each (which has happened), and offer 7 pounds, I know perfectly well to expect 7p change.

This is because

$$7 \times (100 - 1) = 7 \times 100 + 7 \times (-1),$$

which is just Rule 5 again, but with c negative.

So...why does $(-1) \times (-1) = +1$?

While I have heard all sorts of bizarre 'explanations' for this, the short answer is: *it follows directly from the rules.*

Fig. 40 A 'wonder' of mathematics?

Note, first, that any number multiplied by zero is zero. In particular, then,

$$-1 \times 0 = 0.$$

So

$$-1 \times [1 + (-1)] = 0.$$

On using the distributive law, Rule 5, we therefore find that

$$-1 \times 1 + (-1) \times (-1) = 0.$$

So

$$-1 + (-1) \times (-1) = 0,$$

and by adding 1 to both sides we at last obtain the result we have been seeking.

And the consequences are truly enormous.

10

It's a Square World

When we multiply any number *x* by *itself*, we write the result as x^2 and call it '*x* squared'.

The physical world is full of such numbers. When air streams past a wing, for instance, the lift on the wing is proportional to the *square* of the speed of the stream U. So, if an aircraft flies twice as fast, then, other things being equal, the lift goes up not by a factor of 2, but by a factor of 4.

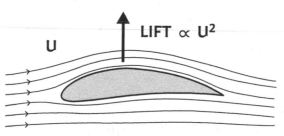

Fig. 41 Airflow past a wing

A more classical example is provided by Galileo's famous experiment with a ball rolling down an inclined plane (Fig. 42).

For, as he discovered, the distance travelled is proportional to the *square* of the elapsed time t.

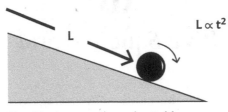

Fig. 42 An experiment by Galileo

Yet the whole idea of 'x squared' goes back much further still, and to purely geometrical origins; for if x is a positive number, then x^2 is, indeed, the area of a square with side of length x (Fig. 43).

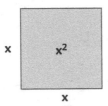

Fig. 43 The area of a square

The importance of being square

We showed in Chapter 9 that $(-1) \times (-1) = +1$, and, in the same way, *any* negative number, when multiplied by itself, gives a positive number.

In consequence, then,

$$x^2 \geq 0$$

whether x is positive or negative.

Or, to put it more geometrically, if we plot a graph of x^2 against x, the curve never dips below the x-axis (Fig. 44).

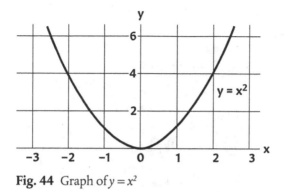

Fig. 44 Graph of $y = x^2$

So, whenever we have a quantity x^2, we immediately know its *sign*, whether x itself is positive or negative.

And, strange as it may seem, it is almost impossible to overestimate the importance of this particular result. My mind truly boggles at the number of mathematical proofs that rely, at some point, on precisely this idea.

It even played a key part in my own first exciting discovery in mathematical research, over 50 years ago.

Square roots

A *square root* of a positive number x is, essentially, the opposite idea: a number which, when multiplied by itself, gives x.

And, in view of what we have just seen, every positive number has two square roots: one positive and one negative. The square roots of 9, for instance, are 3 and −3.

The symbol $\sqrt{x}$ denotes the *positive* square root of the positive number x. Thus $\sqrt{9} = 3$.

"A wonderful square root. Let us hope
it can be used for the good of mankind."
© Sidney Harris

Fig. 45 A wonderful square root?

Square roots, too, occur all over physics, and one of the most well-known examples—the simple pendulum—is associated, again, with Galileo (Fig. 46).

Longer pendulums oscillate to and fro more slowly, and the time T for one complete oscillation increases in proportion to the *square root* of the length ℓ.

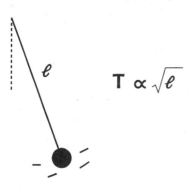

$$T \propto \sqrt{\ell}$$

Fig. 46 The simple pendulum

And we can, if we wish, confirm this by experiment quite easily.

Just take a length of string with a small bob on one end, set it swinging, and count every time it performs *half* a complete oscillation by reaching one end or other of its swing.

Then, while still counting, shorten the string by a factor of 4.

When you set the pendulum swinging again it should then perform—quite convincingly—a complete to-and-fro oscillation in time with your count.

11
Algebra in Action

One of the most important results in the whole of algebra is shown in Fig. 47.

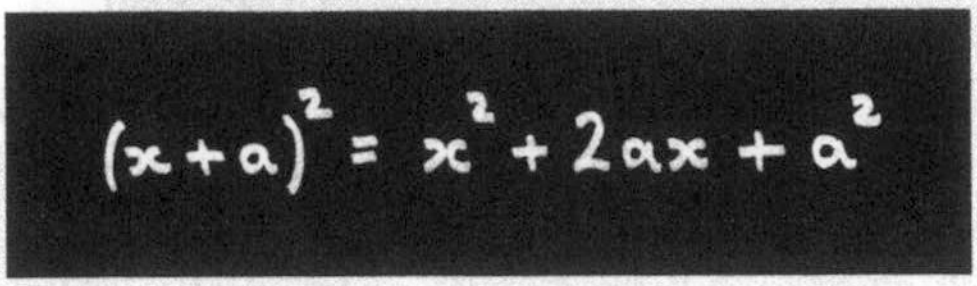

$$(x+a)^2 = x^2 + 2ax + a^2$$

Fig. 47 Algebra at its best

This holds for *any* values of x and a, but I would like to offer, first, a 'picture proof' for the case when both x and a are positive.

A 'picture proof'

In Fig. 48 we have a square of side $x + a$, and therefore area $(x + a)^2$. But it is composed of two smaller squares, with area x^2 and a^2, together with two rectangles, each of area ax.

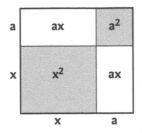

$$(x+a)^2 = x^2 + 2ax + a^2$$

Fig. 48 A picture proof

This leads directly, then, to the result in question, at least for positive values of x and a.

Proof by algebra

This establishes the result in Fig. 47 for *all* x and a, positive or negative, and begins as follows:

$$(x+a)^2 = (x+a)(x+a)$$
$$= (x+a)x + (x+a)a$$

by Rule 5 from p. 50.

The alternative form of Rule 5 (twice) then allows us to rewrite this new expression as

$$x^2 + ax + xa + a^2,$$

and because $xa = ax$ the result then follows:

$$(x+a)^2 = x^2 + 2ax + a^2.$$

And I would now like to use this, without further ado, to prove the most famous theorem in the whole of mathematics.

Pythagoras' theorem

It is well known, I think, that if the two shorter sides of a right-angled triangle are of length 3 and 4, then the longest side—or hypotenuse—is of length 5 (Fig. 49).

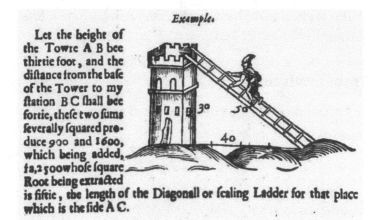

Example.

Let the height of the Towre A B bee thirtie foot, and the distance from the base of the Tower to my station B C shall bee fortie, these two sums severally squared produce 900 and 1600, which being added, fa,2500 whose square Root being extracted is fiftie, the length of the Diagonall or scaling Ladder for that place which is the side A C.

Fig. 49 The 3–4–5 special case of Pythagoras' theorem, from John Babington's *Treatise of Geometrie* (1635)

While this *is* an example of Pythagoras' theorem, the theorem itself is much more general, and provides an unexpectedly simple relationship between the three sides *a*, *b*, *c* of *any* right-angled triangle (Fig. 50).

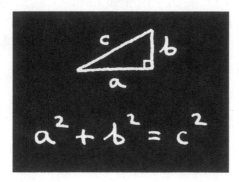

Fig. 50 Pythagoras' theorem

And to prove it, all we need is Fig. 51, where we have placed four copies of the triangle in a square of side $a + b$, leaving a square of side c in the middle.

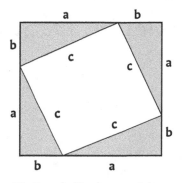

Fig. 51 Proof of Pythagoras' theorem

Now, the area of the large square is

$$(a+b)^2 = a^2 + 2ab + b^2.$$

But it is also equal to c^2 plus the total area of the four triangles. And each of those has area $\frac{1}{2}ab$ (being half that of a rectangle with sides a and b).

So the area of the large square is also

$$c^2 + 2ab$$

and therefore

$$a^2 + b^2 = c^2.$$

12

'Compleating the Square'

One of the oldest problems in mathematics is to be found on a Babylonian clay tablet, dating from about 1800 BC.

This features a square which has been extended by one unit in two directions (Fig. 52), so that the total area is now 120 square units. And the question is: *how long was the side of the square?*

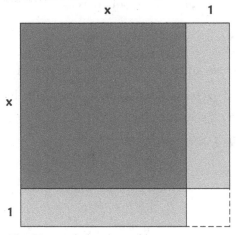

Fig. 52 A Babylonian problem

The Babylonian solution is to add one further unit square, so 'completing' a new square with area 121. *Its* side must therefore be $\sqrt{121}$, which is 11. So the side of the original square must have been 10.

Another view

Today, we usually take a more algebraic approach to the whole idea of 'completing the square'.

If we let x denote the side of the original square in Fig. 52, then its area will be x^2, each of the extensions will have area x, and the Babylonian problem becomes equivalent to solving the *quadratic equation*

$$x^2 + 2x = 120.$$

And we can do this by adding 1 to both sides, which converts the equation into

$$(x+1)^2 = 121,$$

not because of some geometry, but because

$$x^2 + 2x + 1 = (x+1)^2$$

for all x, this being a special case of Fig. 47 in Chapter 11 (with $a = 1$).

It then follows that $x + 1 = \pm 11$. So, as the original problem needs a positive solution, x must be 10, as before.

Completing the square

More generally, whenever we find ourselves with x^2 *plus some 'known' multiple of* x:

$$x^2 + kx$$

it can be helpful to take *half* the coefficient of x, square that, and add it on:

$$x^2 + kx + \left(\frac{k}{2}\right)^2,$$

because this produces

$$\left(x + \frac{k}{2}\right)^2,$$

once again by Fig. 47 in Chapter 11 (with $a = k/2$).

Let $xx + 14\,x$ be compleated.
Here half of 14 is 7, this fquared is 49; fo that $xx + 14x$ when compleated is $xx + 14x + 49$.
Nov. Very eafy indeed, and very pretty.

Fig. 53 From *The Young Algebraist's Companion* (1751)

'Completing the square' in this way can be quite a neat trick, and in the second edition of Daniel Fenning's *Young Algebraist's Companion* (1751), the student (*Nov.*) is at first rather pleased with the whole idea (Fig. 53).

Sadly, however, the next example

Let $xx + 5x$ be compleated

causes more difficulty, for *Nov.* is reduced to mumbling

I am at a Loss at present, indeed

at which point the teacher more or less loses the will to live, and reminds *Nov.* that half of 5 is $^5/_2$.

But what's it all *for*?

In fairness to *Nov.*, it often takes time for any powerful new technique in mathematics to settle in the mind. And it also helps to have some idea, at least, of what the new technique might be *for*.

Firstly, 'completing the square' can be used to solve *any* quadratic equation, in just the same way that we solved $x^2 + 2x = 120$ a moment ago.

But it can also be of great value in maximization problems, where we are trying to make something as large as possible.

And, as we will see in Chapter 17, it can even help with playing snooker!

After x

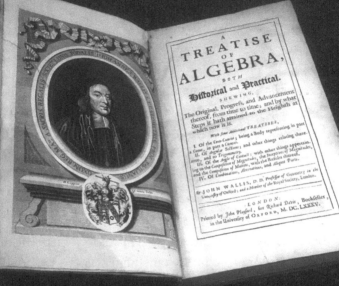

1685   **John Wallis's** *Treatise* was a major exposition of algebra.

But the subject as a whole was still viewed with great suspicion. The philosopher Thomas Hobbes (1588–1679) famously described Wallis's earlier writings on algebra as 'a scab of symbols'.

1775

By the 4th edition of **Thomas Simpson's** *Treatise of Algebra* things are looking more 'familiar', even if the language is not …

Let $\begin{Bmatrix} x + y = 13 \\ x + z = 14 \\ y + z = 15 \end{Bmatrix}$; to find x, y, and z.

By subtracting the first equation from the second (in order to exterminate x) we have $z - y = 1$; to which the third equation being added, y will likewise be exterminated, there coming out $2z = 16$, or $z = 8$: whence y ($= z - 1$) $= 7$; and x ($= 13 - y$) $= 6$.

EXTERMINATE !

13
Slices of Pi

There's nothing quite like *infinity* for bringing mathematics fully to life.

And this happens, believe it or not, as soon as we start thinking seriously about the geometry of a circle.

Most of us begin, I think, by regarding it as intuitively 'obvious' that the circumference of a circle is proportional to its diameter, and this allows us to define the special number

$$\pi = \frac{\text{circumference}}{\text{diameter}},$$

which will be the same for all circles, regardless of their size.

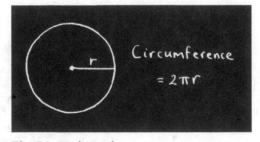

Fig. 54 Circles and π

As the diameter is twice the radius r, the well-known formula in Fig. 54 then follows immediately, and is, more or less, just a restatement of what we actually mean by the number π.

But what about the numerical *value* of π? How are we to determine that?

Measuring π

The most obvious method, I think, is direct physical measurement, and I recently attempted this at home, in my own kitchen.

As my circular object, I chose a gramophone record of French accordion music. I taped a pointer to it, and then rolled it carefully, through one full turn, along the kitchen floor (Fig. 55).

In this way, and with the aid of a steel ruler, I found the circumference to be 94.6 cm.

I then divided by the diameter (30.1 cm) to obtain

$$\pi \approx 3.143.$$

Fig. 55 Measuring π

While I was quite pleased with this, Archimedes did even better, over 2,000 years ago, by proving rigorously that

$$3\frac{10}{71} < \pi < 3\frac{1}{7},$$

which corresponds to

$$3.1408\ldots < \pi < 3.1428\ldots$$

In fact, Archimedes' upper bound of 22/7 was still being used as a 'practical' approximation to π in my early school-days in the 1950s.

The area of a circle

Fig. 56 shows another famous formula involving π, but proving it turns out to be a rather subtle matter.

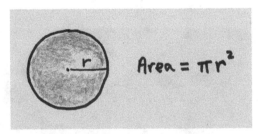

Fig. 56 The area of a circle

After all, we can, if we wish, use *circumference* $= 2\pi r$ to remove π from the scene altogether, and rewrite this new proposition as

$$\text{Area of circle} = r \times \frac{1}{2}\ \text{circumference.}$$

Why on earth should *this* be true?

Infinity enters the picture

To see why, imagine, first, slicing the circle into eight cake-like pieces and reassembling them (Fig. 57a).

The result is vaguely rectangular. The top and bottom have length exactly equal to half the circumference, but are not quite straight, while the left- and right-hand ends are exactly of length r, but not quite vertical.

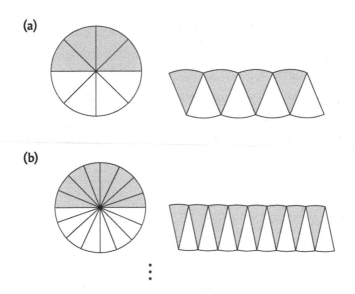

(a)

(b)

Fig. 57 Closer and closer...

Suppose, however, that we now use twice as many pieces, each half as thick (Fig. 57b). The result is clearly much closer to a rectangle.

Moreover—and I hope you agree with this—by continuing *indefinitely* in this way we can make the reassembled figure *as close as we like* to a rectangle with length $\frac{1}{2}$ × circumference and height *r*.

And as the circumference is $2\pi r$, this is why the area of the circle itself must be πr^2.

* * *

Establishing a mathematical idea in this way, by *gradually edging closer and closer to it*, in—so to speak—an infinite number of steps, is a major feature of more advanced mathematics.

But if we are tempted to bypass that whole procedure and simply claim that we would have an exact rectangle with an infinite number of infinitely thin pieces, we are playing with fire.

For, as we will see in Chapter 27, infinity in mathematics needs to be handled very carefully indeed.

In the meantime, however, let us continue to be a little daring.

14
The Golden Ratio

The golden ratio

$$\frac{1+\sqrt{5}}{2} = 1.618\ldots$$

is a special number in mathematics, and crops up in all sorts of unlikely places.

But many people meet it for the first time, I think, through the famous *Fibonacci sequence*

$$1, 1, 2, 3, 5, 8, 13, 21, 34 \ldots$$

in which each term (after the second) is the sum of the previous two.

Fig. 58 Fibonacci and the rabbits

This first arose in a rather dubious model of rabbit-breeding in Fibonacci's *Liber Abaci* of 1202.

And one major feature of this sequence is that the ratio of successive terms gets closer and closer to the golden ratio as the sequence goes on (Fig. 59).

1/1	1.000
2/1	2.000
3/2	1.500
5/3	1.667
8/5	1.600
13/8	1.625
21/13	1.615
34/21	1.619
55/34	1.617

Fig. 59 Closer and closer…

In this sense, then, 'infinity strikes again', because the ratio of successive terms is never *exactly* the golden ratio; it just gets closer and closer to it as the sequence goes on.

A quadratic equation

If we accept that the ratio of successive terms does indeed approach one particular number x as the sequence goes on, it is quite easy to see why x must be the golden ratio.

For three successive terms will then be ever more closely in proportion $1 : x : x^2$ as the sequence goes on, yet the defining property of the Fibonacci sequence is that each term (beyond the second) is the sum of the previous two.

So if this special number x exists at all, it must satisfy the quadratic equation

$$x^2 = x + 1.$$

And to solve that, all we have to do is rewrite it as

$$x^2 - x = 1$$

and then 'complete the square', in the manner of Chapter 12, by adding $\left(-\dfrac{1}{2}\right)^2 = \dfrac{1}{4}$ to each side.

This produces

$$\left(x - \frac{1}{2}\right)^2 = \frac{5}{4},$$

so that $x - \dfrac{1}{2}$ must be $\pm\sqrt{5}\,/2$. But only one of these leads to a positive value for x:

$$x = \frac{1 + \sqrt{5}}{2},$$

and this is, indeed, the golden ratio.

15
Proof by Chocolate

One of the most important ways in which infinity enters mathematics is through the whole idea of an *infinite series*, such as

$$\frac{1}{4}+\frac{1}{16}+\frac{1}{64}+\cdots=\frac{1}{3}$$

As the dots suggest, the series on the left-hand side goes on *for ever*, with each new term four times smaller than the previous one.

Yet the 'sum' is finite, $\frac{1}{3}$, and I would like to now prove this in a slightly unconventional way...using chocolate.

Fig. 60 A special offer

Imagine, if you will, that you are a shopkeeper, and selling chocolate bars on the 'special offer' in Fig. 60.

Now, each bar consists of 1 unit of actual chocolate *and* 1 *coupon*. And a customer with 4 coupons can claim 1 bar free.

The key to proving our result turns out to be the following question:

What is one coupon worth, in terms of actual chocolate?

In search of an infinite series

Well, 4 coupons are worth 1 bar, and therefore 1 coupon is, in a sense, worth $\frac{1}{4}$ bar.

So 1 coupon is worth

$$\frac{1}{4} \text{ [chocolate + coupon]},$$

where 'chocolate' here is just my shorthand for the amount of actual chocolate in one bar.

Now, the coupon in brackets is *itself* worth $\frac{1}{4}$ bar, so 1 coupon can also be said to be worth

$$\frac{1}{4}\left[\text{ chocolate} + \frac{1}{4}\left(\text{ chocolate + coupon} \right) \right].$$

And if we continue like this *for ever*, we find that 1 coupon is worth

$$\left(\frac{1}{4} + \frac{1}{16} + \frac{1}{64} + \cdots \right) \text{ units of actual chocolate.}$$

To prove our result, then, we just need to prove that 1 coupon is also worth 1/3 unit of actual chocolate.

Fig. 61 Proof by chocolate

In search of 1/3

To put it another way, we need to prove that 3 coupons are worth 1 unit of actual chocolate.

And to do this, imagine that I have saved up 3 coupons, come to your shop, and say:

> Good morning. I should like to buy a bar of chocolate.
> I intend to eat the chocolate straight away, in your shop, and I will pay you when I have eaten it.

Your first thought will doubtless be 'another crazy mathematician!', but, being of generous spirit, you give me a bar of chocolate.

I open it, and eat the chocolate.

Next, I take out the coupon that's inside, and add it to the 3 that I have already, so that *I now have 4 coupons!*

I then present these to you, saying: 'Here are 4 coupons. I claim a bar of chocolate!'

At which point, you say: 'You've got a short memory, haven't you? I've just *given* you a bar of chocolate!'

Somewhat embarrassed, I reply: 'Of course you have. How silly of me. So you don't owe me anything, and I don't owe you anything, either. Our business is complete. Thank you very much. Good morning.'

Now, I arrived with 3 coupons and no chocolate, and went away with no coupons and 1 unit of chocolate. So 1 coupon is worth $\frac{1}{3}$ unit of chocolate, and therefore

$$\frac{1}{4} + \frac{1}{16} + \frac{1}{64} + \dots = \frac{1}{3}$$

* * *

And whatever we may think of such an unusual proof, the result itself contains a very important idea: it is possible for *infinitely many positive numbers to have a finite sum.*

Playing with Infinity

The **AWARD** for the largest number of $\sqrt{2}$s in a single formula **GOES TO...**

Francois Viète (1593)

$$\frac{2}{\pi} = \frac{\sqrt{2}}{2} \times \frac{\sqrt{2+\sqrt{2}}}{2} \times \frac{\sqrt{2+\sqrt{2+\sqrt{2}}}}{2} \times \cdots$$

Enter 1 into a calculator

    DO
        Add 1
        Take square root
    LOOP

$$\sqrt{1+\sqrt{1+\sqrt{1+\cdots}}} = \text{Golden Ratio}$$

But <u>WHY?</u> (Ans. p. 158)

16
The Puzzled Farmer

Some of the most attractive problems in mathematics come from trying to make something as large (or as small) as possible.

And it makes sense, I think, to start with the simplest example, even though it is almost as fanciful as the bath-filling of Chapter 2.

So imagine, if you can, that you are a farmer with 4 km of fencing, and you want to create a rectangular field *of maximum area*.

How should you do it?

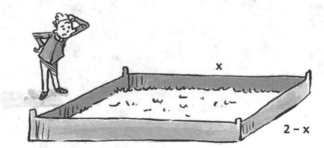

Fig. 62 The puzzled farmer

Well, if we let two sides be of length x, the other two will be of length $2 - x$ (Fig. 62), and the area will therefore be

$$A = 2x - x^2.$$

Now, it isn't immediately obvious how to maximize this. By increasing x, for instance, we can increase the first term, but this will also increase the size of the second term, which is negative.

Suppose, however, that we rewrite the area as

$$A = -(x^2 - 2x).$$

If we now take the expression in brackets and 'complete the square', in the manner of p. 64, we get

$$x^2 - 2x + 1 = (x - 1)^2,$$

which allows us to rewrite A in the form

$$A = 1 - (x - 1)^2.$$

As $(x-1)^2 \geqslant 0$, it follows at once that A can never exceed 1, and only achieves that maximum value when $x = 1$, in which case $2 - x$ is also 1, and the field is a *square*.

While the result itself may not be particularly surprising, the problem as a whole does indicate well how 'completing the square' can help with *optimization* in mathematics.

And it also paves the way for something rather more striking.

An imaginative solution

Suppose now that the situation is slightly different, and that the farmer can make use of a straight stone wall, so that he only has to fence three of the sides (Fig. 63).

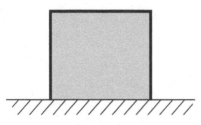

Fig. 63 A slightly different problem

Now we could, of course, go about this new problem in exactly the same way (p. 159)—but there is a cunning alternative.

Imagine, if you will, that as our farmer tries out rectangles of different proportions on one side of the wall, there is *another farmer on the other side,* copying his every move and doing exactly the same thing (Fig. 64).

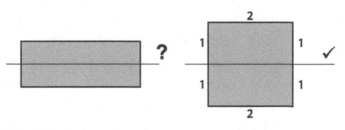

Fig. 64 An imaginative solution

Now, whether they realize it or not, they are then—as a pair—trying to maximize the area of a rectangle which is totally enclosed by a given length of fencing (namely, twice the original).

And, as we have just seen, that is done by making the rectangle a square.

So, with a wall present, an aspect ratio of 2:1 produces the biggest area.

And however implausible the first farmer may have been, the second one was entirely imaginary—in my view, at least—and invented purely to help solve the problem.

17
Mathematics and Snooker

I once played snooker in front of a packed audience at the Crucible Theatre in Sheffield.

All right—it wasn't the World Championships. And it wasn't on a full-size table, either (Fig. 65).

It was actually part of a show for teenagers called 'Maths Inspiration', and I was trying to demonstrate how 'completing the square' can explain why one snooker shot is more difficult than another.

Fig. 65 Setting up at the Crucible Theatre, February 2015. Rob Eastaway, the Director of 'Maths Inspiration', is on the right.

Suppose that luck is with us, and that the cue ball, the red ball, and the pocket are perfectly in line (Fig. 66).

Then a natural question to ask is: which position of the red ball corresponds to the most difficult shot?

In other words, given the distance D and the radius r of a snooker ball, which value of x in Fig. 66 gives the greatest amplification of any small initial error with the cue?

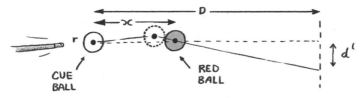

Fig. 66 Mathematics and snooker

Well, suppose first that if the red ball were *not there* the cue ball would miss the centre of the pocket by a very small distance d.

Then a little geometry shows that the centre of the red ball will miss the centre of the pocket by a small amount d', where

$$\frac{d'}{d} = \frac{(D-x)(x-2r)}{2Dr}$$

(see Notes, p. 160).

This is therefore a sort of 'amplification factor' of the initial error, caused by the presence of the red ball.

Now, D and r are given constants, so when we multiply out the brackets we get a quadratic expression in x. And this means that we can use the same trick of 'completing the square', just as in the farmer's problem of Chapter 16.

While the algebra here is a little more messy (p. 161), the outcome is that the amplification factor d'/d is greatest when

$$x = \frac{1}{2}D + r.$$

Now, in practice, r is usually much less than $\frac{1}{2}D$, so the most difficult shot—as every experienced player knows—is when the red ball is *just over halfway to the pocket*.

Moreover, if we calculate the amplification factor itself in that worst case we find it to be

$$\frac{d'}{d} \simeq \frac{D}{8r}.$$

And with a full-length diagonal shot on a full-size snooker table this works out at about 10, explaining why that particular shot is a tricky one, even for top professional players.

18
The Wicked Schoolteacher

Some years ago I met a much-loved primary school teacher who occasionally tormented her little pupils in the following way.

After introducing the idea of square numbers, she would observe that if you take, say, $4 \times 4 = 16$, and multiply instead the two whole numbers 'either side', you get $3 \times 5 = 15$, which is one less.

She would then give one or two other examples, such as the one in Fig. 67.

Fig. 67 A mathematical challenge

Finally, she would offer a chocolate bar to the first person who could find an example where the second product is one *more* than the first, rather than one less.

This would result in a vast amount of (increasingly frenzied) practice in multiplication, largely because the second product is *never* one more than the first; it is always one less.

The reason is that

$$(x-1)(x+1) = x^2 - 1$$

for *any* positive whole number x.

And this, in turn, is only one special case of something far more general.

Another general result

The result in Fig. 68 is true for any numbers x and y, positive or negative.

$$x^2 - y^2 = (x - y)(x + y)$$

Fig. 68 Another example of algebra at its best

And to prove it, all we have to do, once again, is apply the rules of Chapter 9:

$$\begin{aligned}
(x-y)(x+y) &= (x-y)x + (x-y)y \\
&= x^2 - yx + xy - y^2 \\
&= x^2 - y^2.
\end{aligned}$$

In my experience, at least, this result is rather less powerful that the one in Fig. 47, but it still has its moments.

Fig. 69 Lewis Carroll (self-portrait)

And one of them occurred in a curious letter of 20 May 1885 penned by Lewis Carroll.

He was writing to a young boy who was studying algebra, and posed the following problem:

If x and y are each equal to '1', it is plain that $2(x^2 - y^2) = 0$ and also that $5(x - y) = 0$. Hence $2(x^2 - y^2) = 5(x - y)$.

Now divide each side of this equation by $(x - y)$.

Then $2(x + y) = 5$.

But $(x + y) = (1 + 1)$, i.e. $= 2$

So that $2 \times 2 = 5$.

Ever since this painful fact has been forced upon me, I have not slept more than 8 hours a night, and have not been able to eat more than 3 meals a day.

I trust you will pity me and will kindly explain the difficulty.

The puzzled farmer (again!)

The result in Fig. 68 also provides a neat alternative approach to the farmer's problem in Chapter 16.

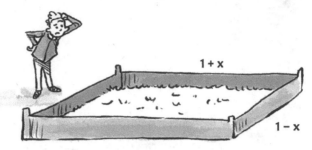

Fig. 70 The puzzled farmer

The idea is to *start* with a square of side 1 and let x denote the *extension* to two of the sides (Fig. 70).

As the total length of fencing is fixed (at 4), the other two sides will have to shorten by the same amount x, so the area will then be

$$A = (1+x)(1-x)$$

$$= 1 - x^2,$$

which plainly takes its maximum value when $x = 0$.

So, once again, square is best.

And, in many ways, this particular view of things is really just *The Wicked Schoolteacher*, all over again!

Extreme Situations

When light is refracted at the plane boundary between two media it takes the path of **LEAST TIME** between any two given points A and B.

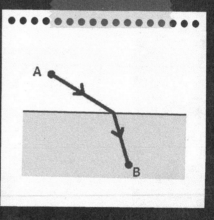

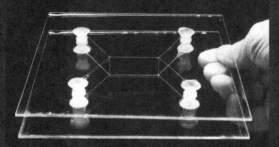

This soap film has found the **SHORTEST PATH** linking all four pins.

Strange Powers

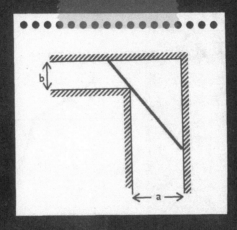

The longest ladder that can get round the corner has length

$$L = \left(a^{2/3} + b^{2/3}\right)^{3/2}$$

($x^{3/2}$ denotes the square root of x cubed, and $x^{2/3}$ the cube root of x squared.)

19

Trains, Boats, and Planes

Problems involving *motion* can offer some good opportunities for elegant mathematics with simple materials.

The key idea is that if something is moving at *constant speed*, then the distance travelled is proportional to the time taken, and the speed itself is just the ratio of the two:

$$\text{speed} = \frac{\text{distance travelled}}{\text{time taken}}$$

Fig. 71 Trains, boats, and planes

And to show this idea in action, I would like to begin with some puzzles, each of which has some interesting mathematical features.

Passing trains

A train travels at constant speed from A to B in T_A minutes. Starting at the same time, another train travels at constant speed from B to A in T_B minutes.

At what time (after the start) will they meet?

Fig. 72 Passing trains

Well, it makes sense, surely, to begin by letting T denote this 'meeting time'.

Now, as the first train is travelling at constant speed, the distance travelled will be proportional to the time taken, so at time T it will have completed a fraction T/T_A of its journey.

In the same way, the second train will have completed a fraction T/T_B.

But the two fractions must evidently add up to 1, and dividing by T then gives

$$\frac{1}{T} = \frac{1}{T_A} + \frac{1}{T_B}.$$

While I prefer the elegance of this form of the answer, we can rewrite it, if we wish, as

$$T = \frac{T_A T_B}{T_A + T_B}.$$

After the event

An interesting variation on this problem appeared in Thomas Simpson's *Treatise of Algebra* of 1745 (Fig. 73).

PROBLEM XLVI.

A Traveller sets out from B to go to C, at the same time as another sets out from C to go to B ; they both travel uniformly, and in such proportion to each other, that the former, four Hours after their meeting, arrives at C, and the latter at B, in nine Hours after : Now the Question is, to find in how many Hours each Person performed his Journey.

B———————————D———————C

Fig. 73 From Simpson's *Treatise of Algebra* (1745)

There we are given not the journey times, but the journey *completion* times T_a, T_b *after the meeting*, and the time at which they meet is then

$$T = \sqrt{T_a T_b}$$

(see Notes, p. 162).

Once again, the meeting time is pleasantly (though also necessarily) symmetric with respect to A and B.

A windy flight

An aeroplane flies in a straight line from A to B, then back in a straight line from B to A. It travels with constant speed and there is no wind.

Suppose now that during the whole journey, at the same constant engine speed, a constant wind blows from A to B.

Will the total travel time be the same as before, or greater, or less?

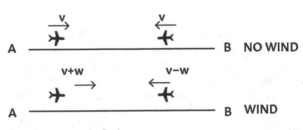

Fig. 74 A windy flight

At first sight, perhaps, it will be the same—the wind will speed things up on the way out, slow things down on the way back, and the two effects will perhaps cancel out.

But we can demolish this idea immediately with a very general checking procedure in mathematics which consists in *taking things to extremes.*

Suppose, then, that the wind is *really* strong—in fact almost (but not quite) equal to the engine speed. Then the journey time out will be almost halved—which is good. But the trouble is, the journey back will be absolutely horrific, because the aeroplane will be making appallingly slow progress against the wind.

In fact, the total journey time is *always* longer when there is a constant wind.

And we can prove it as follows.

Let D be the distance between A and B, and v the aeroplane speed relative to still air.

Then the total travel time with no wind is

$$T_1 = \frac{2D}{v}.$$

But if the wind speed is w, the total travel time is

$$T_2 = \frac{D}{v+w} + \frac{D}{v-w}$$

$$= \frac{D(v-w) + D(v+w)}{(v+w)(v-w)}$$

and with the help of the key result in Fig. 68 this simplifies to

$$T_2 = \frac{2Dv}{v^2 - w^2}.$$

So

$$\frac{T_2}{T_1} = \frac{v^2}{v^2 - w^2},$$

which is always greater than 1, and very large indeed if w is only slightly less than v, as we observed earlier.

The missing hat

A man rows his boat downstream on a river which flows at 2 miles per hour. As he passes a bridge he hears a splash but doesn't realize that his hat has fallen out of the boat and is now floating on the water.

Fig. 75 The missing hat

After an hour, he notices that his hat is missing and remembers the splash under the bridge. His rowing speed in still water is 3 miles per hour.

How long will he need to row back upstream to collect his hat?

* * *

I cannot begin to describe the mess I once made of trying to solve this particular problem, largely by looking at it the wrong way.

It's true that, at the time, I was lazing under a tree in my garden, on a lovely summer afternoon, after a delicious Sunday lunch with lots of wine.

But, even so...

(Answer: p. 162.)

20
I've Seen that Before, Somewhere...

At the risk of appearing somewhat obsessed with the subject, I should like to return for a moment to *bath-filling*.

Suppose that, in Fig. 76, tap A fills the bath in time T_A, while tap B fills it in time T_B, with each filling at a constant rate.

Then, together, they will fill it in time T, where

$$\frac{1}{T} = \frac{1}{T_A} + \frac{1}{T_B}.$$

Fig. 76 Bath-filling revisited

This is startlingly reminiscent of the 'passing trains' problem of Chapter 19. Yet, with a little thought, it is not really surprising.

After all, in each case there are two agencies trying to complete a certain task at constant rates, and it doesn't really matter whether that task is filling a bath or covering the full distance between two stations.

And in the case of the trains, the time T when they meet is precisely the time at which they *jointly* cover that full distance.

Yet connections between different parts of mathematics can sometimes be rather more subtle and intriguing.

A crossed-ladders problem

In geometry, the idea of *similar triangles* is a powerful one.

These are triangles which have exactly the same *shape*, even though they may be of quite different size. In particular, then, they share the same three angles.

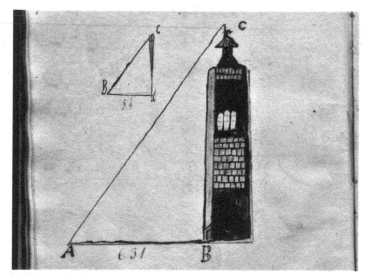

Fig. 77 Finding the height of a tower by 'similar triangles', in William Botham's exercise book of 1819

And the key result is that such triangles have all three sides *in the same proportion*.

The classic example, shown in Fig. 77, is to find the height of a tower by using a short vertical pole, for their heights will be in the same proportion as the lengths of their shadows.

A slightly more exotic problem dates back to AD 850, and an old textbook by the Indian mathematician Mahavira.

Two ladders are propped against the walls of an alley (Fig. 78), with the heights *a* and *b* given, and the problem is to determine the height *h* at which the ladders cross.

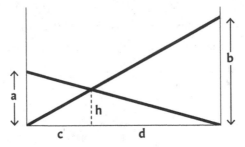

Fig. 78 The crossed-ladder problem

Once again, similar triangles come to the rescue, just as in the 'tower problem'. For, by one pair:

$$\frac{h}{b} = \frac{c}{c+d}$$

and by another:

$$\frac{h}{a} = \frac{d}{c+d}.$$

Adding, and dividing by h, then gives

$$\frac{1}{h} = \frac{1}{a} + \frac{1}{b}.$$

Now, the connection between *this* problem and the 'passing trains' one is, at first sight, rather more mysterious.

Yet there is an explanation.

Motion pictures

It can sometimes be helpful to represent motion at constant speed on a distance–time graph (Fig. 79). And the slope of the line then represents the speed of the object.

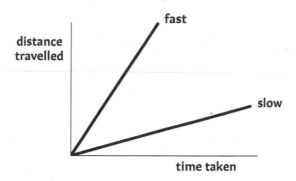

Fig. 79 Distance–time graphs

Suppose, then, that we take the 'passing trains' problem of p. 96 and plot what the two trains are doing on a distance–time graph (Fig. 80).

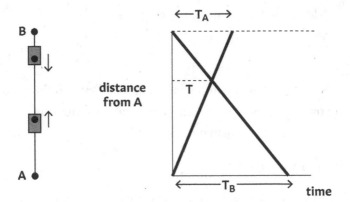

Fig. 80 A fresh look at the 'passing trains' problem

Then, looked at this way, the 'passing trains' problem is, in effect, just the crossed-ladders problem of Fig. 78 *turned through 90°*!

And surprising *connections* like this, between different aspects of mathematics, provide some of the deepest pleasures in the subject, especially at a more advanced level.

21

An Apple Falls...

In the summer of 1666, Isaac Newton saw an apple fall in his garden, and promptly invented the theory of gravity.

That, at least, is how the story goes, and it brings us to the whole subject of *dynamics*, which is about how—and why—things change with time.

For, in sharp contrast to our earlier examples of motion, the speed of the apple changes all the time as it falls.

In fact, if there is no air resistance, the speed at time t after the start is given by

$$v = gt,$$

so that v increases in proportion to t.

Fig. 81 Newton and the apple

The constant of proportionality,

$$g \approx 9.81 \, \text{ms}^{-2},$$

is a measure of how quickly v increases with time, and represents the downward *acceleration*.

The acceleration itself is caused by a downward force F on the apple due to gravity, and if m denotes the mass of the apple, then $F = mg$ (Fig. 82).

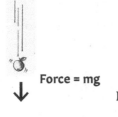

Force = mg

Fig. 82 The force on the apple

The reason for this is Newton's famous law of motion, usually stated today as

Force = Mass × Acceleration.

Before we go any further, however, we should note that 'acceleration' here does not have the same meaning that it often has in everyday life.

When travelling in a car, for instance, we tend to think of acceleration as rate of change of speed, without regard to the direction of motion.

But this is, in fact, mathematically and scientifically inaccurate. Acceleration isn't rate of change of speed; it's rate of change of *velocity*, and velocity is speed *with direction*.

So, even if an object is moving at constant speed, it will have a non-zero acceleration if the direction of motion is changing. And this acceleration will, itself, have a certain direction at any given moment, which may not be intuitively obvious.

Circular motion

Take motion in a circle, for instance (Fig. 83). Even if the speed v is a constant, the velocity is changing in direction all the time, and it turns out that, in consequence, there is an acceleration v^2/r towards the centre of the circle, where r is the radius.

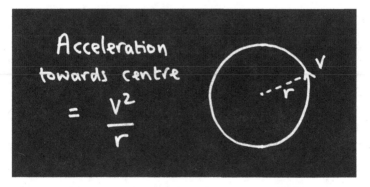

Fig. 83 Circular motion

In case this should seem surprising at all, note that, by Newton's law of motion, this acceleration requires a *force*

$$F = mv^2/r$$

towards the centre, where m is the mass of the object, and this, surely, is in keeping with common experience.

For if we whirl a stone in a sling, for instance, we know that we need to apply a radially inward force on the stone to stop it simply flying off at a tangent (Fig. 84).

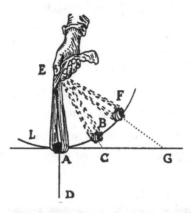

Fig. 84 A stone being whirled in a sling, from Descartes' *Principles of Philosophy* (1644)

It is notable, too, that this force mv^2/r increases with v and also increases as r gets smaller, again in keeping with common experience.

And when all these ideas were applied to the motion of the planets, in the seventeenth century, they led to one of the greatest discoveries in the history of science.

As it happens, however, I have chosen to apply them here to something rather more frivolous...

22
Rollercoaster Mathematics

One of the highlights of a modern rollercoaster ride is 'loop-the-loop', where you travel, for a short time, upside-down.

Fig. 85 shows a toy model, where a ball accelerates down a steeply inclined plane, enters a circular loop, and hasn't got quite enough initial speed to reach the top.

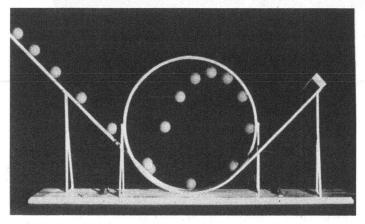

Fig. 85 Not quite loop-the-loop (*Stroboscopic image*)

Yet equivalent loops in theme parks are typically *not* circular, and to see why we need, first, to briefly revisit the falling apple of Chapter 21.

The apple revisited

We have noted already that the downward velocity of the apple is $v = gt$.

As the speed is changing all the time, the familiar idea of *distance = speed × time* does not apply, and it turns out that the distance h fallen in time t is given instead by Fig. 86.

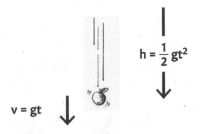

$$h = \frac{1}{2}gt^2$$

$$v = gt$$

Fig. 86 The falling apple

If we eliminate t between the two equations there, we find that $v^2 = 2gh$, and if we then multiply both sides by $\frac{1}{2}m$, where m is the mass of the apple, we obtain something even more significant:

$$\frac{1}{2}mv^2 = mgh.$$

For this is an example of *conservation of energy*; $\frac{1}{2}mv^2$ is the kinetic energy that the apple gains in its fall (from an initial state of rest), and mgh is the gravitational potential energy that it loses by falling a height h.

And one of the key aspects of kinetic energy, in particular, is that to calculate $\frac{1}{2}mv^2$ we only need to know the speed of the motion; its *direction* doesn't matter.

This means, then, that we can apply these ideas at once to our rollercoaster problem.

Rollercoaster dynamics

Suppose that we have a circular loop, radius r, and that a car enters the loop with speed v_B at the bottom and slows to speed v_T at the top (Fig. 87).

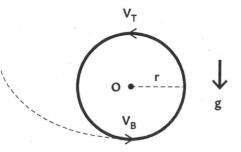

Fig. 87 Rollercoaster dynamics

In going from the bottom to the top, the height of the car increases by $2r$, so equating the potential energy gained to the kinetic energy lost gives

$$\frac{1}{2}mv_B^2 - \frac{1}{2}mv_T^2 = 2mgr.$$

And, somewhat remarkably, if we divide by $\frac{1}{2}mr$ we obtain a relation between the *accelerations* towards the centre O:

$$\frac{v_B^2}{r} = \frac{v_T^2}{r} + 4g.$$

In particular, then, the acceleration towards O at the very start of the loop must be *at least* 4g in order that the car can get to the top.

Now, I have explained everything so far, according to the laws of motion, from the viewpoint of a stationary observer, looking on at events with their feet firmly on the ground.

But a passenger sitting in the car itself will rationalize things quite differently. If their mass is M, they will experience a centrifugal force *downwards* at the start of the loop of at least 4Mg, in addition to their normal weight of Mg.

And the situation will be even worse if the car is not secured to the track, or if the passenger is not strapped into it, for they will then need $v_T^2/r > g$, to counteract the downward pull of gravity at the *top* of the loop and stop them falling out. At the bottom of the loop they will then experience a total downward force of at least 6Mg—or six times their own body weight.

A force of that size would be enough to render most people unconscious, and that is why loop-the-loop rollercoasters have, in practice, a different design that specifically avoids such extreme accelerations.

23
The Electric Guitar Revisited

When a guitar string vibrates in its 'fundamental mode', the frequency of the resulting note is given by the formula in Fig. 88.

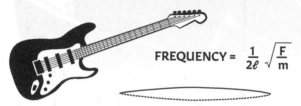

$$\text{FREQUENCY} = \frac{1}{2\ell} \sqrt{\frac{F}{m}}$$

Fig. 88 Vibrations of a guitar string. Here F denotes the tension force in the string, l its length, and m its mass per unit length.

And while it is impossible to establish that formula using 'simple materials' alone, we can get remarkably close!

The secret is to just look carefully at the *dimensions* of the various quantities involved.

The method of dimensions

In any problem of pure mechanics, the physical dimension of each quantity must be some combination of mass (M), length (L), and time (T).

Velocity, for instance, has dimension of L/T. being measurable in metres per second.

Acceleration has dimension of L/T^2. (The acceleration g due to gravity, for instance, is 9.81 m/s^2.)

Force has dimension ML/T^2, essentially because of Newton's law of motion: *Force = Mass × Acceleration*.

Dimensions	
Mass	M
Length	L
Time	T
Velocity	L/T
Acceleration	L/T^2
Force	ML/T^2

Fig. 89 Dimensions

And these ideas can often provide very simple but effective checks on progress in mechanics.

Circular motion

In Chapter 21, for instance, I claimed that when an object moves at speed v round a circle of radius r it has an acceleration v^2/r towards the centre.

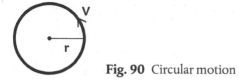

Fig. 90 Circular motion

While the derivation of this result is beyond the scope of this book, we can check very easily that at least its dimensions are correct.

For v^2 has dimension L^2/T^2, and r, being a length, has dimension L. So the dimensions of v^2/r are L/T^2, and these are, indeed, the dimensions of acceleration.

The simple pendulum

As a second example, consider a pendulum consisting of just a bob on the end of a string of length ℓ.

Then the time for one small to-and-fro oscillation turns out to be independent of the mass of the bob and given by the formula in Fig. 91.

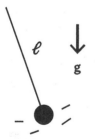

$$\text{OSCILLATION TIME} = 2\pi \sqrt{\frac{\ell}{g}}$$

Fig. 91 The simple pendulum

Now, g denotes the acceleration of a freely falling body under gravity, with dimension L/T^2. So, when we divide the length ℓ by g we get something with dimension T^2. And after taking the square root we end up with something with dimensions of time, just as we should.

Yet, however satisfactory all this may be, the method of dimensions offers, in general, much more than just a check.

The guitar revisited

The physical quantities in play here are shown in Fig. 92, and we want to combine them to find a formula for the vibration frequency.

		Dimension
String length	ℓ	L
Mass per unit length	m	M/L
Tension force	F	ML/T^2

Fig. 92 The guitar revisited

Note that m here denotes not the mass of the string (which would be proportional to ℓ) but the mass of the string *per unit length*, which will be some fixed quantity for the string in question.

So, when we press down on different frets the only thing that will change in the frequency formula is the string length ℓ.

Now, the frequency itself is the number of vibrations *per unit time*, so it has dimension $1/T$.

And there is only one way to get such a quantity from ℓ, m, and F.

First, in order to keep M 'out of the picture', F and m can only enter the formula through the combination F/m, with dimension L^2/T^2.

And the only way of combining this with the string length ℓ to get something with dimension $1/T$ is to take the square root of F/m and then divide by ℓ.

In this way, we learn *from dimensional considerations alone* that the vibration frequency of the string must be proportional to

$$\frac{1}{\ell}\sqrt{\frac{F}{m}},$$

where the constant of proportionality is some pure number (in fact $\frac{1}{2}$), having no dimensions at all.

In particular, then, the vibration frequency must be *inversely proportional to the string length ℓ*.

And in this way, the method of dimensions explains, at a stroke, why pressing down on a higher fret results in a higher note.

A Journey through the Earth?

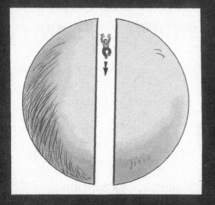

If the Earth were a solid sphere of uniform density, then, in the absence of friction, you would fall right through a hole through its centre in 42 minutes.

More remarkably still, this gravitational journey time would be the same for *all* straight tunnels through the Earth.

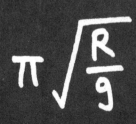

The journey time of 42 minutes comes from this formula, where R is the radius of the Earth and g the acceleration due to gravity *at the Earth's surface.*

$$\pi \sqrt{\dfrac{R}{g}}$$

24
The Domino Effect

There is a famous story about the legendary mathematician Carl Gauss as a young schoolboy in the 1780s.

One day, in a desperate attempt to keep him occupied, his teacher told him to add up all the whole numbers from 1 to 100.

Yet, in no time at all, Gauss had the answer.

His 'trick' was to observe that if you write down the desired sum *backwards*, and then *add*, you get 100 little columns, each with a total of 101. The answer is therefore half of 10,100, which is 5050.

Fig. 93 Some clever addition

Moreover, with the help of a little algebra, we can immediately generalize this to the result in Fig. 94, because reversing and adding then gives n columns, each with a total of $n + 1$.

$$1 + 2 + 3 + \cdots + n = \tfrac{1}{2}n(n+1)$$

Fig. 94 A generalisation

In truth, however, the real reason I have introduced this particular result is so that I can show you a different method of proof altogether.

Proof by induction

We want to prove that the result in Fig. 94 holds for all positive whole numbers n.

Suppose, then, that we happen to know *one particular* value of n—which I will call N—for which it is true, so that

$$\text{Sum of N terms} = \frac{1}{2} N(N + 1).$$

As the $(N + 1)$th term for the series under consideration is just $N + 1$, it would then follow at once that

$$\text{Sum of } N + 1 \text{ terms} = \frac{1}{2} N(N + 1) + (N + 1),$$

and the remarkable thing about the right-hand side is that it can be rewritten as

$$\frac{1}{2} (N + 1) (N + 2).$$

For this is just the original formula $\frac{1}{2} N (N + 1)$ *but with $N + 1$ in place of N!*

Fig. 95 Proof by induction is often compared with knocking down a row of dominoes

We have shown, in other words, that *if* the formula $\frac{1}{2}n(n+1)$ in Fig. 94 'works' for one particular whole number *n*, *then it will also work for the next one*.

Now, it certainly works when $n = 1$, because both sides of the equation in Fig. 94 are then simply 1. So it must also work when $n = 2$. And now that we know it works for $n = 2$, it must also work for $n = 3$, and so on.

And in this very different way, then, we learn (again) that

$$1 + 2 + \ldots + n = \frac{1}{2}n(n+1)$$

for all positive whole numbers *n*.

This whole idea of proof by induction is a powerful one in mathematics, and thought to date from Francesco Maurolico's *Arithmeticorum Libri Duo* of 1575, where it was used to prove that the sum of the first n odd numbers is n^2 (Fig. 96).

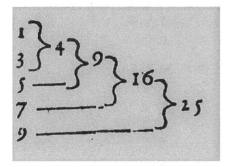

Fig. 96 From Maurolico's *Arithmeticorum Libri Duo* of 1575

For what it's worth, I have always rather liked it, if only for its slight air of 'living dangerously'.

After all, unless you are very careful, you can find yourself assuming, right at the beginning, the very thing you're trying to prove!

Some other examples

As the sum of the first N whole numbers can be established more easily by Gauss's neat trick, it doesn't *quite* show the method of proof by induction at its best.

$$1^2 + 2^2 + \ldots + N^2 = \frac{1}{6} N(N+1)(2N+1)$$

$$1^3 + 2^3 + \ldots + N^3 = \frac{1}{4} N^2 (N+1)^2$$

Fig. 97 Two more results which lend themselves well to proof by induction

If we consider instead, however, the sum of the first N *squares*—or even *cubes*—the situation is quite different, for they do not lend themselves to the same trick at all (Fig. 97). And some readers might even like to try the method of induction on them for themselves (see *Notes*, p.163).

Curiously, perhaps, establishing the sum of the cubes is actually easier, algebraically, than the sum of the squares!

A two-colour theorem

I cannot resist ending this chapter with an attractive but slightly frivolous opportunity for induction which arises from the map-colouring problem of Chapter 7.

For in the *very* special case when the boundaries between regions are formed simply by n straight lines in the plane, with no more than two lines meeting at any one point, only *two* colours are needed.

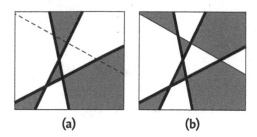

(a)   (b)

Fig. 98 Inductive map-colouring

To prove this, suppose that we know *one particular* value of *n* for which two colours are sufficient.

Then, when we add one more line—dotted in Fig. 98a—the colours will match in an unacceptable way all along that line, but things will still be 'okay' everywhere else.

And we can plainly fix that problem, while leaving everything else still 'okay', if we *flip all the colours on one side of the new line* (Fig. 98b).

It follows, then, that if two colours are enough for one particular value of *n*, they will also be enough for the next value of *n*.

And two colours are plainly enough when $n = 1$.

25
Real or Imaginary?

There is always a certain excitement involved in solving a quadratic equation.

And it is caused by one simple question:

Will the solutions be real, or not?

To see what I mean, consider, if you will, the general quadratic equation in Fig. 99. Here, a, b, and c are 'known' quantities (with a non-zero), and the problem—as ever—is to find x.

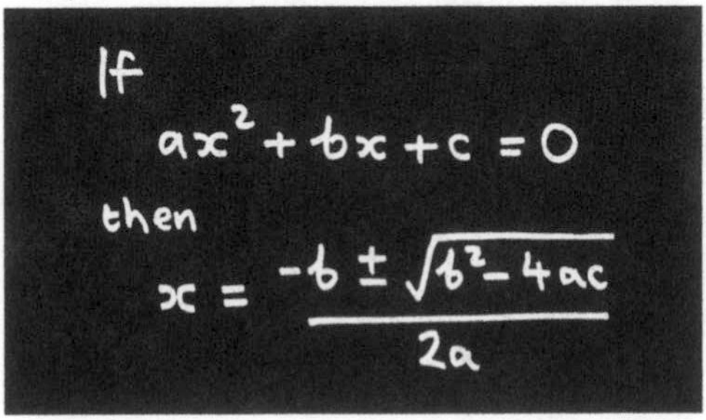

$$\text{If}$$
$$ax^2 + bx + c = 0$$
$$\text{then}$$
$$x = \frac{-b \pm \sqrt{b^2 - 4ac}}{2a}$$

Fig. 99 The general quadratic equation and its solution

Happily, we can do this quite easily, just by dividing by a and then 'completing the square' (see Notes, p. 164). Yet a glance at the solution shows that x will only be *real* if

$$b^2 \geqslant 4ac,$$

for otherwise we end up trying to take the square root of a negative number.

And in scientific and engineering applications, this particular aspect of quadratic equations is often the most important one.

Here's an example…

The spinning top

How does a spinning top manage to avoid falling over?

In particular, is there some critical rotation speed needed to make the upright position stable?

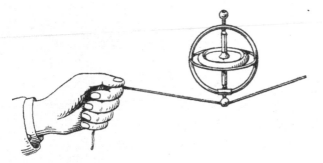

Fig. 100 How does the spinning top stay upright? (From H. Crabtree's *Spinning Tops and Gyroscopic Motion*, Longmans, 1909.)

This is, in truth, a fairly sophisticated problem in mechanics, requiring a great deal of advanced mathematics. Yet it all comes down *in the end* to a quadratic equation.

And for the simple model in Fig. 101, with just a thin circular disc on a (weightless) axle, the quadratic equation in question turns out to be:

$$\frac{5}{4}x^2 - \pi R x + \frac{g}{\ell} = 0.$$

Here, g denotes—as usual—the acceleration of a freely falling object under gravity, and R is the rotation rate of the disc about its axle, in revolutions per second.

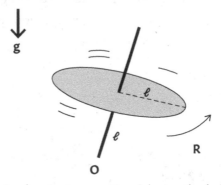

Fig. 101 A simple spinning top. Here ℓ denotes both the radius of the disc and its distance from the fixed point O.

The number x itself is related to the wobble of the spinning top when it is disturbed slightly from its purely upright position. And if x is real, then the top is stable, but otherwise it isn't.

The condition '$b^2 > 4ac$', derived from Fig. 99, tells us, then, that the top will be stable in its upright position if

$$R > \frac{1}{\pi}\sqrt{\frac{5g}{\ell}},$$

but unstable if its rotation rate R falls below this critical value.

This explains, then, what we know from experiment, that the top will only be stable in its upright position if we spin it fast enough. With $\ell = 5$ cm, say, and $g = 981\,\text{cm/s}^2$, the critical value of R works out at about 10 revolutions per second.

And in a similar way, quadratic equations even entered my own mathematical research, on astrophysical fluid dynamics, many years ago.

So, while elementary textbooks of mathematics often focus on real solutions of quadratic equations, and treat others, perhaps, as of relatively little interest, my own experience is the exact opposite: it is often precisely when the solutions *aren't* real that things really begin to liven up!

26
The Square Root of Minus One

What kind of number *is* the square root of −1?

It can't be a 'real' number, because if we square any such number—positive or negative—the result is always positive.

For this reason it is called an 'imaginary' number, and denoted by *i* (Fig. 102).

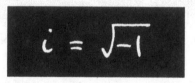

Fig. 102 The square root of minus one

We have seen already how numbers of this kind can emerge from solutions to quadratic equations. Curiously, however, this wasn't how imaginary numbers really entered mathematics.

It was, instead, through *cubic* equations.

The cubic equation

The general solution to the cubic

$$x^3 = px + q$$

is

$$x = \sqrt[3]{\frac{q}{2} + \sqrt{\left(\frac{q}{2}\right)^2 - \left(\frac{p}{3}\right)^3}} + \sqrt[3]{\frac{q}{2} - \sqrt{\left(\frac{q}{2}\right)^2 - \left(\frac{p}{3}\right)^3}},$$

where $\sqrt[3]{}$ denotes a cube root.

It was first published (in word form) by Girolamo Cardano in his *Ars Magna* of 1545, though he acknowledged that the first discoverer had been Scipione del Ferro, a mathematics lecturer in Bologna, in the 1520s.

Fig. 103 Girolamo Cardano (1501–1576)

Cardano also gave credit to Niccolo Tartaglia, from whom he had prised some of the secrets of the cubic in 1539. An enormous row then ensued, because Cardano had promised Tartaglia never to publish his discoveries, and, more bizarrely still,

> to note them down in code, so that after my death, no one will be able to understand them.

In any event, while the general solution of the cubic equation was a great mathematical achievement, actually *imple-
menting* it can be a bit tricky.

Real or imaginary?

The particular case

$$x^3 = 15x + 4$$

illustrates the problem well.

Note, first, that there is clearly a real solution, $x = 4$, as this
makes both sides of the equation equal to 64.

Yet Cardano's general solution gives, in this case,

$$x = \sqrt[3]{2+11i} + \sqrt[3]{2-11i},$$

where $i = \sqrt{-1}$.

How on earth, then, can we get $x = 4$ out of *this*?

It fell to another Italian scholar, Raffaele Bombelli, to show
in his *L'Algebra* of 1572 that it *can* be done, but only by taking
imaginary numbers 'seriously', and subjecting them to all the
usual rules of algebra, replacing i^2 by -1 wherever it occurs.

In particular, he noticed the following. If we take $2 + i$ and
multiply it by itself we get

$$(2 + i)^2 = 4 + 4i + i^2$$
$$= 3 + 4i.$$

And if we now multiply by $2 + i$ *again*, we get

$$(3 + 4i)(2 + i) = 6 + 11i + 4i^2$$
$$= 2 + 11i.$$

L'ALGEBRA
OPERA
Di Rafael Bombelli da Bologna
Diuisa in tre Libri.

Con la quale ciascuno da se potrà venire in perfetta
cognitione della teorica dell'Arimetica.

Con vna Tauola copiosa delle materie, che
in essa si contengono.

posta hora in luce à beneficio delli studiosi di
detta professione.

IN BOLOGNA,
Per Giouanni Rossi. MDLXXIX.
Con licenza de' Superiori.

Fig. 104 Title page of a 1579 edition of Bombelli's *L'Algebra*

So

$$(2+i)^3 = 2+11i,$$

and, in a similar way,

$$(2-i)^3 = 2-11i.$$

This allowed Bombelli to interpret Cardano's general solution

$$x = \sqrt[3]{2+11i} + \sqrt[3]{2-11i}$$

as

$$x = 2+i+(2-i)$$
$$= 4.$$

And it was in this way, then, by resolving a mystery concerning *cubic* equations, that imaginary numbers really entered the mathematical landscape.

More Playing with Infinity

$$\frac{1}{4} + \frac{1}{4^2} + \frac{1}{4^3} + \ldots = \frac{1}{3}$$

1736

Leonhard Euler discovers this extraordinary infinite series for π:

$$1 + \frac{1}{2^2} + \frac{1}{3^2} + \cdots = \frac{\pi^2}{6}$$

What, then, is

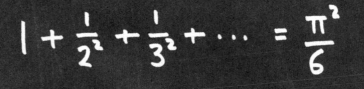

$$1 + \frac{1}{3^2} + \frac{1}{5^2} + \frac{1}{7^2} + \cdots$$

(Ans. p. 165)

27
Inspector Riemann Investigates...

It was a cold winter's morning in 1956, and Inspector Riemann of Scotland Yard had been called in to investigate a strange burglary.

A large amount of money—£346,573—had mysteriously disappeared from a locked safe, yet there were no signs of a break-in at all.

With his keen intellect and finely chiselled features, Riemann lost no time in asking the owner about the detailed contents of the safe.

Fig. 105 Inspector Riemann investigates

'Lots of credit and debit notes', said the man.

'How many?' asked Riemann.

'*Infinitely* many', said the man.

'That's a lot', said Riemann, beginning to wonder just how well he was going to cope with this particular case.

'The strange thing is...', continued the man, '...nothing is actually missing.'

'You mean they're all still there?'

'Well', said the man, 'I checked the contents of the safe last night, before going home. There was a credit note for £1 million, then a debit note for £ $\frac{1}{2}$ million, a credit for £ $\frac{1}{3}$ million, a debit for £ $\frac{1}{4}$ million, and so on.'

'And what does all that come to?' asked Riemann.

'Well', said the man, 'in millions, it comes to

$$1-\frac{1}{2}+\frac{1}{3}-\frac{1}{4}+\frac{1}{5}-\frac{1}{6}+\frac{1}{7}-\frac{1}{8}\ldots,$$

and that's 0.693147...'

'You mean £693,147?'

'Yes', said the man. 'The thing is, when I came in this morning I thought I'd better check it again, by counting all the credit and debit notes *but in a different order*. So after each credit note I counted *two* of the debit ones.'

'You mean

$$\left(1-\frac{1}{2}-\frac{1}{4}\right)+\left(\frac{1}{3}-\frac{1}{6}-\frac{1}{8}\right)+\left(\frac{1}{5}-\frac{1}{10}-\frac{1}{12}\right)+\ldots?'$$

said Riemann.

Fig. 106 A mysterious disappearance

'Yes, exactly', said the man. 'And that's the same as

$$\frac{1}{2}-\frac{1}{4}+\frac{1}{6}-\frac{1}{8}+\frac{1}{10}-\frac{1}{12}+\dots,$$

'And what does that come to?'

'Well', said the man, '*that's* the same as

$$\frac{1}{2}\left(1-\frac{1}{2}+\frac{1}{3}-\frac{1}{4}+\frac{1}{5}-\frac{1}{6}+\dots\right),$$

which is only *half the original amount!'*

'So £346,573 has disappeared?' said Riemann.

'Exactly', said the man. 'And what I want to know is: *where's it gone?'*

28

Infinite Danger

Actually, there wasn't a 'robbery' at all.

But there *is* something a bit mysterious going on, and to understand it we need to take a more cautious look at infinite series in general.

The first one we met, using chocolate, was

$$\frac{1}{4} + \frac{1}{16} + \frac{1}{64} + \ldots = \frac{1}{3}.$$

We have also seen a 'picture proof' of the same result (p. 136), and Fig. 107 shows another.

Fig. 107 Another 'picture-proof'

In truth, however, all these proofs are a little informal and risky.

A safer and more cautious approach with any infinite series is always to consider first the sum S_n of *just the first n terms*, and then examine carefully what happens to that as n gets ever larger.

In this particular case, then,

$$S_n = \frac{1}{4} + \frac{1}{4^2} + \ldots + \frac{1}{4^n},$$

and multiplying by 4 gives

$$4S_n = 1 + \frac{1}{4} + \ldots \frac{1}{4^{n-1}}.$$

If we then subtract S_n from this, there is a spectacular amount of cancellation, leaving

$$3S_n = 1 - \frac{1}{4^n}$$

and, plainly, as n gets larger and larger, $1/4^n$ gets smaller and smaller, and S_n gets closer and closer to $1/3$.

More significantly still, we can make S_n *as close to 1/3 as we like* for all sufficiently large values of n. And that is, more precisely, what the three dots in

$$\frac{1}{4} + \frac{1}{4^2} + \frac{1}{4^3} + \ldots = \frac{1}{3}$$

really mean.

And in case you should wonder why I am suddenly being so cautious about all this, here's the answer.

Infinity again

In 1350, the French scholar Nicole Oresme studied the infinite series

$$1+\frac{1}{2}+\frac{1}{3}+\frac{1}{4}+\cdots$$

and showed that it has *no finite sum*, even though the individual terms get smaller and smaller as the series goes on.

Fig. 108 Nicole Oresme (1320–1382)

He began by grouping the terms, after the first, in the following way:

$$\frac{1}{2}$$

$$\frac{1}{3}+\frac{1}{4},$$

$$\frac{1}{5}+\frac{1}{6}+\frac{1}{7}+\frac{1}{8}$$

$$\vdots$$

so that each new group has twice as many terms as the previous one.

Oresme then observed that $\frac{1}{3} + \frac{1}{4}$ is greater than $\frac{1}{4} + \frac{1}{4} = \frac{1}{2}$, that the next group is greater than $\frac{1}{8} + \frac{1}{8} + \frac{1}{8} + \frac{1}{8} = \frac{1}{2}$, that the one after that is greater than $8 \times \frac{1}{16} = \frac{1}{2}$, and so on, for ever.

With Oresme's series, then, we can make the sum S_n of the first n terms *as big as we like* by taking n large enough.

And, believe it or not, this provides the key to the mysterious 'robbery' in Chapter 27.

What robbery?

As mentioned already, there wasn't really a robbery at all.

And Riemann wasn't a police inspector at Scotland Yard, either. He was a famous German mathematician in the nineteenth century, who proved that the infinite series in Fig. 109

Fig. 109 A puzzling series

really does have different 'sums' *depending on the order in which we add up the terms!*

The simplest way of seeing this is, perhaps, to take the extreme case in which we decide to count all the positive terms first:

$$1 + \frac{1}{3} + \frac{1}{5} + \frac{1}{7} + \ldots$$

The trouble with this is that, like Oresme's series, it has no finite sum; we can make the running total as large as we like by counting enough terms.

And, in the same way, the running total of the negative terms

$$-\frac{1}{2} - \frac{1}{4} - \frac{1}{6} - \ldots$$

heads towards *minus* infinity—as is evident when we realize that it is $-\frac{1}{2}$ times Oresme's original series.

Suddenly, then, it is rather less surprising that if we combine positive and negative terms in our running total, the final outcome will depend rather critically on just how we do it.

Riemann showed, in fact, that we can make the running total of this series converge to *any value we like* by taking the positive and negative terms in a sufficiently cunning order.

No wonder infinity in mathematics has to be handled rather carefully!

29
1 + 1 = 2 to the Rescue!

It isn't often, I think, that 1 + 1 = 2 provides the key to a tricky problem in mathematics.

To see an example, suppose we want to make a soup can of given volume using as little material as possible (Fig. 110).

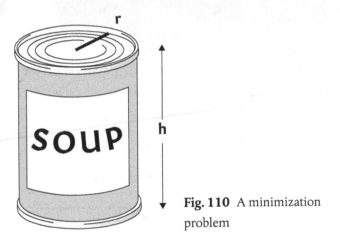

Fig. 110 A minimization problem

The top and bottom will each have circumference $2\pi r$ and area πr^2, and on multiplying by the height h in each case we find that the area of the curved surface is $2\pi rh$ and the volume is $\pi r^2 h$.

We want, then, to minimize the total surface area $A = 2\pi r^2 + 2\pi rh$ while keeping the volume $V = \pi r^2 h$ fixed.

And if we eliminate h, this amounts to finding the value of r which minimizes

$$A = 2\pi r^2 + \frac{2V}{r}$$

for a given, fixed value of V.

Now, it is not immediately obvious, I think, how to do this. Reducing r, for instance, will make the first term, $2\pi r^2$, smaller, but it will make the second term *bigger*.

Fortunately, help is at hand from a powerful new idea.

The AM-GM inequality

This major result is shown in Fig. 111.

On the left-hand side we have the *arithmetic mean* (AM) of n positive numbers. This is what is commonly called their 'average', obtained by simply adding the numbers and dividing by n.

THE AM-GM INEQUALITY

$$\frac{1}{n}\left(x_1 + x_2 + \ldots + x_n\right) \geq \sqrt[n]{x_1 x_2 \ldots x_n}$$

Fig. 111 The AM-GM inequality for n positive numbers $x_1 \ldots x_n$

On the right-hand side we have a different kind of 'average', called the *geometric mean* (GM), obtained instead by *multiplying* the numbers together and then taking the nth root.

Remarkably, the AM is always greater than the GM *unless the n numbers are all equal*, in which case the AM and GM are themselves equal (see Notes, p. 166).

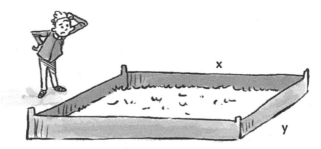

Fig. 112 The puzzled farmer, for the last time!

Suppose, for instance, that the puzzled farmer of Chapter 16 is again constructing a rectangular field, but wants instead to find the shortest length of fencing that will enclose a given area A. Then, with reference to Fig. 112, this amounts to minimizing $2(x + y)$ subject to the condition that $xy = A$.

Now, the $n = 2$ version of the AM-GM inequality tells us at once that

$$\frac{1}{2}(x+y) \geq \sqrt{xy},$$

with equality only when $x = y$.

Notably, the right-hand side of the inequality is a *known constant*, equal to $\sqrt{A}$, for our particular problem. So $x + y$ is never less than $2\sqrt{A}$, and only reaches that minimum value when $x = y$, in which case the field is—yet again—a square.

And essentially the same approach can be used to solve our soup can problem—though there's a twist…

An unexpected twist

In this case, we start with some fixed, given value of the volume V, and our problem is to choose r so that the surface area

$$A = 2\pi r^2 + \frac{2V}{r}$$

is as small as possible.

As A involves the sum of two terms it is only natural to try the $n = 2$ version of the AM-GM inequality again, but this time it is virtually useless for our purposes, because the right-hand side turns out to be proportional to $\sqrt{r}$, and we have no idea at this stage how large or small r might be.

The 'trick' then, in this particular case, is to use $1 + 1 = 2$ to split the term $2V/r$ into two equal parts, for the $n = 3$ version of the AM-GM inequality then tells us that

$$\frac{1}{3}\left(2\pi r^2 + \frac{V}{r} + \frac{V}{r}\right) \geqslant \sqrt[3]{2\pi r^2 \cdot \frac{V}{r} \cdot \frac{V}{r}}$$

Crucially, the cube root on the right-hand side is now a 'known' *constant*, independent of r, because we have organized things so that the various factors of r cancel.

So A is never less than three times the cube root of $2\pi V^2$, and it only achieves that minimum value when all three terms on the left-hand side are equal, i.e. when

$$2\pi r^2 = \frac{V}{r}.$$

On appealing once more to $V = \pi r^2 h$, this means that $2r = h$, so that the minimum surface area is achieved when *the diameter of the can is equal to its height.*

And, in a sense, the key to it all was $1 + 1 = 2$.

30
And Finally…

We have come a long way since the faintly absurd problems of Chapter 2, featuring A, B, and C.

In particular, we have seen many aspects of *algebra* at work, often to express some general idea in mathematics, or to act as a bridge between mathematics and the physical world.

We have seen, too, some distinctive mathematical reasoning, such as proof by contradiction and the method of dimensions, to say nothing of a somewhat bizarre 'proof by chocolate'.

Finally, we have engaged with *infinity* from time to time, and this provides a pathway, in fact, to much of higher mathematics, including calculus.

A long way, indeed, from A, B, and C.

Yet, as it happens, I would like to end with a short footnote on that very topic.

* * *

For I had always casually assumed that the three characters A, B, and C themselves were inventions of some long-forgotten nineteenth-century textbook.

I therefore got something of a shock recently when I acquired a copy of Isaac Newton's *Universal Arithmetick* (1728).

The title denotes what we would call algebra, and the book itself is a translation (from Latin) of lectures that Newton gave in Cambridge between 1673 and 1683.

And one of his examples (Fig. 113) has A, B, and C hard at work some 200 years earlier than I had always imagined!

Fig. 113 A, B, and C in the 2nd edition of Isaac Newton's *Universal Arithmetick* (1728)

E x a m p l e. Three Workmen can do a Piece of Work in certain Times, *viz.* *A* once in 3 Weeks, *B* thrice in 8 Weeks, and *C* five times in 12 Weeks. It is defired to know in what Time they can finifh it jointly? Here then are the Forces of

Did Newton perhaps even *invent* A, B, and C? While problems of this general kind date back to ancient times, A, B, and C themselves seem to be an entirely different matter.

In any event, Newton's example has one additional and most arresting feature.

For he tells us confidently that A can do the work in 3 weeks, B can do it 3 times in 8 weeks, and C can do it 5 times in 12 weeks.

On close inspection, then, C *is 'winning', yet again!*

Notes

3. THE 1089 TRICK

There is a slight difficulty if the first figure exceeds the last *by only* 1.

Reversing and subtracting then gives 99, and the trick looks doomed to fail.

Yet it can *still* be made to work if we can find an excuse for writing 99 as 099, and the 1922 dollars-and-cents version in Fig. 114 is a nice example.

	$6.73	$9.91	$2.31	$0.01
	3.76	1.99	1.32	1.00
dif.	2.97	7.92	0.99	0.99
	7.92	2.97	9.90	9.90
sum	$10.89	$10.89	$10.89	$10.89

Fig. 114 From Sloane's *Rapid Arithmetic* (1922)

* * *

The most general form of the trick that I know is for the case when the large unit is p times the medium unit and the medium unit is q times the small unit. The final answer is then always $(pq - 1)(q + 1)$ small units.

In particular, when p and q are both 10, this gives 99×11, which is indeed 1089.

* * *

My knowledge of the trick's history comes largely from a conversation with David Singmaster in December 2002, at a Christmas meeting of the British Society for the History of Mathematics.

Further details of David's research into this (and other aspects of 'recreational mathematics') can be found at www.puzzlemuseum. com/singma/singma-index.htm.

6. A MOST UNUSUAL LECTURE

The proof of infinitely many primes in Euclid's *Elements*, Book 9, Proposition 20, is structured rather differently, and is *not* a 'full-blown' proof by contradiction.

The idea, instead, is to take any finite collection of primes, multiply them all together, and add 1. The resulting number must then be either

    (a) prime or
    (b) not prime, in which case it must have prime factors
    which do not belong to the original collection (because
    division by any of the originals leads to a remainder of 1).

Either way, then, some new prime must exist which does not belong to the original collection.

So, no matter how many primes you've got, there will always be some more.

8. PUZZLING MATHEMATICS

A chocolate bar problem

The number of breaks needed is *always* 23, because every time you pick up a piece—no matter what its shape—the number of individual pieces increases by 1 when you break it in one of the ways indicated.

The roll of the dice

The answer is 6.

The most elegant solution I know, which focuses almost ruthlessly on the one thing we want to find, is to imagine the dice at the *end* of the track, and then imagine rolling it *backwards*, to find out where the top face started.

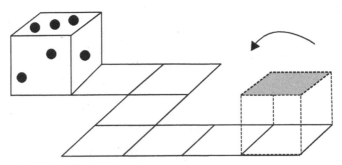

Fig. 115 Rolling backwards…

As we see it, in Fig. 115, the face we want to know about starts on top, then faces left, goes to the bottom, faces right, stays on the right having turned the corner, emerges on top after turning the second corner, and therefore starts facing *left*.

C 'wins' again!

It is simplest, I think, to view the whole problem relative to, say, A, who is then essentially 'fixed', while B moves at 8 − 5 = 3 miles a day and C moves at 10 − 5 = 5 miles a day.

When B has completed N whole revolutions, C will have completed $5N/3$ revolutions. We therefore want the smallest whole number N for which $5N/3$ is also a whole number. This is plainly N = 3.

As B performs one revolution in $73/3$ days. The answer is 73 days.

The mutilated chessboard

For many mathematicians, I think, a problem of the kind 'show that such-and-such is impossible' is, more or less, an open invitation to try proof by contradiction.

Suppose, then, that it *is* possible.

As each domino covers one black square and one white square, the numbers of black and white squares on the 'mutilated' board must be equal.

But they're not, because while they were equal originally, we removed *two black squares* to obtain Fig. 32.

So our original supposition must be wrong.

A four-card puzzle

We need to turn over the first *and last* cards.

For if the first has an odd number on the other side, *or* the last is black on the other side, the rule is false. Otherwise, it is true.

15. PROOF BY CHOCOLATE

I first came across this quirky (and, I think, little-known) proof 60 years ago in a book called *Playing with Infinity* (Bell, 1961) by the Hungarian mathematician Rózsa Péter, though she herself attributed it to her mentor László Kalmár.

Playing with infinity

Assuming that such a (finite) number exists, let

$$x = \sqrt{1 + \sqrt{1 + \sqrt{1 + \cdots}}}$$

Then

$$x^2 = 1 + \sqrt{1 + \sqrt{1 + \cdots}}$$
$$= 1 + x,$$

which is the same quadratic equation that led, on p. 75, to the *Golden Ratio*

$$x = \frac{1 + \sqrt{5}}{2}.$$

* * *

By the same reasoning,

$$\sqrt{2 + \sqrt{2 + \sqrt{2 + \dots}}} = 2,$$

and this is, in part, how Viète's extraordinary infinite product on p. 80 manages to converge to a finite value; successive terms in the product get closer and closer to 1.

16. THE PUZZLED FARMER

With the wall present, and x defined as in Fig. 116, the more standard 'completing the square' solution goes as follows:

$$\begin{aligned} A &= x(4 - 2x) \\ &= -2x^2 + 4x \\ &= -2(x^2 - 2x) \end{aligned}$$

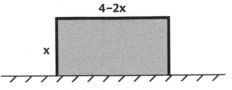

Fig. 116 The field-and-wall problem

Completing the square gives

$$A = -2(x - 1)^2 + 2,$$

so A is a maximum when $x = 1$, in which case $4 - 2x = 2$.

17. MATHEMATICS AND SNOOKER

The formula for d'/d on p. 87 may be obtained as follows.

Let the centres of the cue ball and red ball be initially at A and B, let C be the centre of the cue ball at the moment of impact, and let G be the middle of the pocket.

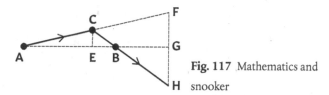

Fig. 117 Mathematics and snooker

Now, triangles AEC and AGF are 'similar', i.e. exactly the same shape. So their various sides must all be in the same proportion, and therefore

$$\frac{CE}{FG} = \frac{AE}{AG}.$$

In the same way

$$\frac{GH}{CE} = \frac{BG}{EB}.$$

Eliminating CE gives

$$\frac{GH}{FG} = \frac{BG.AE}{EB.AG},$$

and on translating all this into the notation of Fig. 66 (p. 87), we have

$$\frac{d'}{d} = \frac{(D-x).AE}{EB.D}.$$

Up to this point, the analysis has been exact—on the assumption that the red ball goes off along the line of centres CB.

But if the initial error is very small, so that CE is very small compared to AB, then EB is almost equal to CB—which is 2r—and AE is almost $(x - 2r)$, whence the result.

* * *

The details of 'completing the square' on p. 87 go as follows:

$$
\begin{aligned}
(D-x)(x-2r) & \\
&= -x^2 + (D+2r)x - 2rD \\
&= -\left[x^2 - (D+2r)x\right] - 2rD \\
&= -\left[x - \left(\frac{D}{2}+r\right)\right]^2 + \left(\frac{D}{2}+r\right)^2 - 2rD
\end{aligned}
$$

So the maximum of $(D - x)(x - 2r)$ occurs when $x = \frac{1}{2}D + r$, and that maximum value is

$$
\left(\frac{D}{2}+r\right)^2 - 2rD = \left(\frac{D}{2}-r\right)^2
$$

18. THE WICKED SCHOOLTEACHER

In Lewis Carroll's problem, $x - y = 0$, because x and y are both equal to 1. So the whole argument is really just a thinly disguised version of the following incorrect 'reasoning':

$$
0 \times 4 = 0 \times 5,
$$

(because both sides are zero)—therefore, *dividing by zero*:

$$
4 = 5!
$$

19. TRAINS, BOATS, AND PLANES

After the event . . .

With reference to this slightly varied problem in which we are given T_a and T_b instead (p. 97), let the passing point be a distance D_A from A and D_B from B.

Then, as the train from A is travelling at constant speed, and takes time T to get to the passing point and time T_a thereafter:

$$\frac{T}{T_a} = \frac{D_A}{D_B}.$$

The same considerations for the train from B give

$$\frac{T}{T_b} = \frac{D_B}{D_A}.$$

So $T^2 = T_a T_b$.

In consequence, the meeting time in Simpson's problem (Fig. 73) is $\sqrt{4 \times 9} = 6$ hours after the start.

The missing hat

The answer is 1 hour, and by far the simplest way of seeing this is to view the whole episode from a frame of reference *moving with the river*.

The water then appears still, the hat *stays where it fell*, and he rows away for 1 hour. So it takes him 1 hour to row back to get it.

(His rowing speed in still water, and the speed of the river, are both irrelevant.)

A journey through the Earth

This all assumes *no friction*. It also assumes (quite wrongly) a completely solid Earth of uniform density.

The acceleration due to gravity within the Earth is then proportional to the distance from the centre, and this plays a key part in the detailed mathematical theory.

If you fall into the hole from a state of rest, you will achieve your maximum speed at the halfway point, then come gently to rest again at the other end—momentarily—before falling back in again (unless somebody grabs you!).

The reason why a journey through a very short tunnel will *still* take 42 minutes is that such a tunnel will be almost perpendicular to the Earth's gravitational field, so the component of gravity *in the direction of the tunnel* will be very small.

24. THE DOMINO EFFECT

The key calculations for Fig. 97 are, for the sum of the cubes:

$$\frac{1}{4}N^2(N+1)^2 + (N+1)^3$$
$$= \frac{1}{4}(N+1)^2\{N^2 + 4(N+1)\}$$
$$= \frac{1}{4}(N+1)^2(N+2)^2,$$

and for the sum of the squares:

$$\frac{1}{6}N(N+1)(2N+1) + (N+1)^2$$
$$= \frac{1}{6}(N+1)\{N(2N+1) + 6(N+1)\}$$
$$= \frac{1}{6}(N+1)(2N^2 + 7N + 6)$$
$$= \frac{1}{6}(N+1)(N+2)(2N+3)$$

25. REAL OR IMAGINARY?

To solve

$$ax^2 + bx + c = 0,$$

first divide by a (which is non-zero):

$$x^2 + \frac{b}{a}x = -\frac{c}{a}$$

then 'complete the square' on the left-hand side, as on p. 64, by adding $\left(b\big/_{2a}\right)^2$ to both sides:

$$\left(x + \frac{b}{2a}\right)^2 = \left(\frac{b}{2a}\right)^2 - \frac{c}{a} = \frac{b^2}{4a^2} - \frac{c}{a} = \frac{b^2 - 4ac}{4a^2}$$

So

$$x + \frac{b}{2a} = \pm\sqrt{\frac{b^2 - 4ac}{4a^2}} = \pm\frac{\sqrt{b^2 - 4ac}}{2a}$$

whence the general solution in Fig. 99 on p. 128.

26. THE SQUARE ROOT OF MINUS ONE

To solve the *cubic* equation

$$x^3 = px + q,$$

first write x as the sum of two parts:

$$x = u + v.$$

Substituting in, and multiplying everything out, gives

$$u^3 + v^3 + 3uv(u+v) = p(u+v) + q.$$

Introducing *two* new variables in this way—rather than just one—gives us an element of choice, and we now choose v so that

$$v = \frac{p}{3u}.$$

This results in a lot of cancellation, leaving

$$u^3 + v^3 = q,$$

and if we then substitute for v we end up with a *quadratic equation* in the variable u^3.

Solving that for u^3, and then using $v^3 = q - u^3$, gives the general solution to the cubic on p. 133.

More playing with infinity

From what we are given (due to Euler):

$$1 + \frac{1}{3^2} + \frac{1}{5^2} + \ldots = \frac{\pi^2}{6} - \left(\frac{1}{2^2} + \frac{1}{4^2} + \frac{1}{6^2} + \ldots \right)$$

But the infinite series on the right-hand side is the same as

$$\frac{1}{4} \left(\frac{1}{1^2} + \frac{1}{2^2} + \frac{1}{3^2} + \ldots \right),$$

i.e. one-quarter of $\pi^2/6$. So

$$1 + \frac{1}{3^2} + \frac{1}{5^2} + \ldots = \frac{\pi^2}{8}.$$

29. 1 + 1 = 2 TO THE RESCUE!

Proof of the AM–GM inequality

Let $x_1, x_2, \ldots, x_n$ be n positive numbers with geometric mean

$$G = \sqrt[n]{x_1 x_2 \ldots x_n}$$

and arithmetic mean

$$A = \frac{1}{n}(x_1 + x_2 + \ldots + x_n).$$

If the n numbers are all equal to X, say, then $A = G = X$.

Suppose now that the n numbers are *not* all equal.

Then some will inevitably be greater than G, and some will be less than G, because their geometric mean is G.

The idea now is to pick one of each, a and b, say, so that $a < G < b$, and *replace* them by the numbers G and ab/G. As the product of the two numbers is the same as before, i.e. ab, this will leave the geometric mean of all n numbers unchanged (and equal to G), but it will *decrease the arithmetic mean*, because

$$a + b > G + \frac{ab}{G}.$$

(This inequality follows directly from $(G - a)(b - G) > 0$.)

By continuing in this way, with the arithmetic mean decreasing further at each step, after at most $n - 1$ steps *all* the n numbers will have been replaced by G, and the final arithmetic mean will therefore also be G.

The *original* arithmetic mean A must therefore have been greater than G, which is what we were trying to prove.

30. AND FINALLY . . .

Newton in fact precedes the A, B, and C example in Fig. 113 with an algebraic treatment *of the general case* (Fig. 118).

PROBLEM VII.

The Forces of several Agents being given, to determine x the Time, wherein they will jointly perform a given Effect d.

Let the Forces of the Agents A, B, C, be supposed, which in the Times e, f, g can produce the Effects a, b, c respectively; and these in the Time x will produce the Effects $\frac{ax}{e}$, $\frac{bx}{f}$, $\frac{cx}{g}$; wherefore is $\frac{ax}{e} + \frac{bx}{f} + \frac{cx}{g} = d$, and by Reduction $x = \dfrac{d}{\frac{a}{e} + \frac{b}{f} + \frac{c}{g}}$.

Fig. 118 From Newton's *Universal Arithmetick* (1728)

The original manuscript (in Latin) can be viewed online at Cambridge University Digital Library, MS Add. 3993. It is partly in Newton's own hand and partly in the hand of his amanuensis Humphrey Newton. It dates from the 1680s, and A, B, and C appear on p. 81.

Further Reading

1089 and All That, by David Acheson (Oxford University Press, 2002).

Alex's Adventures in Numberland, by Alex Bellos (Bloomsbury, 2010).

17 Equations That Changed the World, by Ian Stewart (Profile Books, 2012).

The Joy of X, by Steven Strogatz (Atlantic Books, 2012).

The Mathematical Universe, by William Dunham (Wiley, 1994).

A History of Pi, by Petr Beckmann (Golem Press, 1970).

The Great Mathematicians, by Raymond Flood and Robin Wilson (Arcturus, 2011).

For the history of algebra, I recommend:

Taming the Unknown, by Victor J. Katz and Karen Hunger Parshall (Princeton University Press, 2014).

A Discourse Concerning Algebra, by Jacqueline A. Stedall (Oxford University Press, 2002).

And for the history of algebra in physics:

The Mathematics of Measurement, by John J. Roche (Athlone Press, 1998).

Acknowledgments

I am extremely grateful, as ever, to the staff of Oxford University Press for taking such care in the production of this book, and special thanks are due to Emma Slaughter, Henry Clarke, and Latha Menon.

I would also like to thank Patrick Leger for the cover illustration, and Jon Davis for some of the more exotic internal figures, such as 5, 67, 71, and 105.

Picture Credits

The publisher and author apologize for any errors or omissions in the above list. If contacted they will be pleased to rectify these at the earliest opportunity.

Index

1089 AND ALL THAT

A Journey into Mathematics

David Acheson

978-0-19-959002-5 | Paperback | £8.99

'Every so often an author presents scientific ideas in a new way...Not a page passes without at least one intriguing insight...Anyone who is baffled by mathematics should buy it. My enthusiasm for it knows no bounds.' *Ian Stewart, New Scientist*

'An instant classic...an inspiring little masterpiece.' *Mathematical Association of America*

'Truly inspiring, and a great read.' *Mathematics Teaching*

This extraordinary little book makes mathematics accessible to everyone. From very simple beginnings Acheson takes us on a journey to some deep mathematical ideas. On the way, via Kepler and Newton, he explains what calculus really means, gives a brief history of pi, and introduces us to chaos theory and imaginary numbers. Every short chapter is packed with puzzles and illustrated by world famous cartoonists, making this is one of the most readable and imaginative books on mathematics ever written.

THE WONDER BOOK
OF GEOMETRY

A Mathematical Story

David Acheson

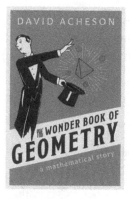

978-0-19-884638-3 | Hardback | £12.99

'David Acheson has set geometry free from the confines of stuffy textbooks and lets loose its potential to surprise and delight. There's a rich and ancient history to be found in these pages, and a future for the field that extends beyond neat (yet elegant) equations.' *BBC Science Focus, Books of the Year*

'Give this to a curious teenager and they will fall in love with geometry.' *Alex Bellos*

How can we be sure that Pythagoras's theorem is really *true*? Why is the 'angle in a semicircle' always 90 degrees? And how can tangents help determine the speed of a bullet?

David Acheson takes the reader on a highly illustrated tour through the history of geometry, from ancient Greece to the present day. He emphasizes throughout elegant deduction and practical applications, and argues that geometry can offer the quickest route to the whole spirit of mathematics at its best. Along the way, we encounter the quirky and the unexpected, meet the great personalities involved, and uncover some of the loveliest surprises in mathematics.

THE CALCULUS STORY

A Mathematical Adventure

David Acheson

978-0-19-880454-3| Hardback | £11.99

Calculus is the key to much of modern science and engineering. It is the mathematical method for the analysis of things that change, and since in the natural world we are surrounded by change, the development of calculus was a huge breakthrough in the history of mathematics. But it is also something of a mathematical adventure, largely because of the way infinity enters at virtually every twist and turn…

In *The Calculus Story* David Acheson presents a wide-ranging picture of calculus and its applications, from ancient Greece right up to the present day. Drawing on their original writings, he introduces the people who helped to build our understanding of calculus. With a step by step treatment, he demonstrates how to start doing calculus, from the very beginning.

EULER'S PIONEERING EQUATION

The most beautiful theorem in mathematics

Robin Wilson

978-0-19-879492-9 | Hardback | £14.99

The story of a supremely elegant equation which connects five of the most important concepts in mathematics

In just seven symbols, Euler's Equation connects five of the most important ideas in mathematics – our counting system; the concept of zero; the irrational number π; the exponential e; and the imaginary number i. Robin Wilson explains how mathematicians arrived at their understanding of each of these – and how Euler brought them all together.

MATH HYSTERIA

Fun and games with mathematics

Ian Stewart

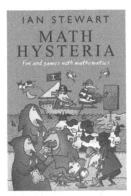

Professor Stewart presents us with a wealth of magical puzzles, each one spun around an amazing tale: Counting the Cattle of the Sun; The Great Drain Robbery; and Preposterous Piratical Predicaments; to name but a few. Along the way, we also meet many curious characters: in short, these stories are engaging, challenging, and lots of fun!

978-0-19-861336-7 | Paperback | £12.99

HOW TO CUT A CAKE

And other mathematical conundrums

Ian Stewart

From the author of Math Hysteria

HOW TO CUT A CAKE

AND OTHER
MATHEMATICAL
CONUNDRUMS

IAN STEWART

978-0-19-920590-5 | Paperback | £11.99

Twenty curious puzzles and fantastical mathematical tales from Professor Ian Stewart, one of the world's most popular and accessible writers on mathematics.

Welcome to Ian Stewart's magical world of mathematics! This is a strange world of never-ending chess games, empires on the moon, furious fireflies, and, of course, disputes over how best to cut a cake. Each quirky tale presents a fascinating mathematical puzzle – challenging, fun, and also introducing the reader to a significant mathematical problem in an engaging and witty way.

COWS IN THE MAZE

And other mathematical explorations

Ian Stewart

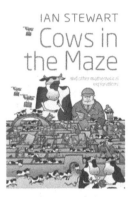

978-0-19-956207-7 | Paperback | £8.99

From the mathematics of mazes, to cones with a twist, and the amazing sphericon – and how to make one – Ian Stewart is back with more mathematical stories and puzzles that are as quirky as they are fascinating, and each from the cutting edge of the world of mathematics.

We find out about the mathematics of time travel, explore the shape of teardrops (which are not tear-drop shaped, but something much, much more strange!), dance with dodecahedra, and play the game of Hex, amongst many more strange and delightful mathematical diversions.

FROM HERE TO INFINITY

Ian Stewart

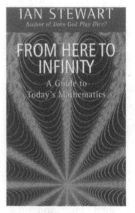

978-0-19-283202-3 | Paperback | £11.99

A retitled and revised edition of Ian Stewart's *The Problem of Mathematics*, this is the perfect guide to today's mathematics. Read about the latest discoveries, including Andrew Wile's amazing proof of Fermat's Last Theorem, the newest advances in knot theory, the Four Colour Theorem, Chaos Theory, and fake four-dimensial spaces. See how simple concepts from probability theory shed light on the National Lottery and tell you how to maximize your winnings. Discover how infinitesimals become respectable, why there are different kinds of infinity, and how to square the circle with the mathematical equivalent of a pair of scissors.